U0294560

RHR

RIVER HAPPINESS REPORT

WORLD RIVER HAPPINESS REPORT 2021 EXECUTIVE SUMMARY

China Institute of Water Resources and Hydropower Research

· Beijing ·

图书在版编目（CIP）数据

世界河流幸福指数报告. 2021 : 执行摘要 = World River Happiness Report 2021 Executive Summary : 英文 / 中国水利水电科学研究院著. -- 北京 : 中国水利水电出版社, 2023.6
ISBN 978-7-5226-1617-9

Ⅰ. ①世… Ⅱ. ①中… Ⅲ. ①河流－生态环境建设－研究报告－世界－2021－英文 Ⅳ. ①X321.2

中国国家版本馆CIP数据核字(2023)第126904号

书名	**World River Happiness Report 2021 Executive Summary**
作者	China Institute of Water Resources and Hydropower Research
出版发行	中国水利水电出版社 (北京市海淀区玉渊潭南路1号D座 100038) 网址: www. waterpub. com. cn E-mail: sales@ mwr. gov. cn 电话: (010) 68545888 (营销中心)
经售	北京科水图书销售有限公司 电话: (010) 68545874、63202643 全国各地新华书店和相关出版物销售网点
排版	中国水利水电出版社微机排版中心
印刷	天津嘉恒印务有限公司
规格	210mm×285mm 16开本 4.5印张 137千字
版次	2023年6月第1版 2023年6月第1次印刷
印数	0001—3300册
定价	**98.00**元

凡购买我社图书，如有缺页、倒页、脱页的，本社营销中心负责调换

Research Team

Director: Kuang Shangfu

Deputy Directors: Peng Jing Wang Jianhua Cao Wenhong

Board Members: Peng Wenqi Jiang Yunzhong Lv Juan Guo Qingchao Li Yinong

Peng Xiang Zhang Jianli Li Lulu Qu Xiaodong Liu Changshun

Technical Leaders: Peng Wenqi Qu Xiaodong Liu Changshun

Team Members :

Wang Shan	Wang Jing	Wang Dandan	Wang Shiyan	Wang Weihao
Mao Yu	Qiu Yaqin	Deng Jun	Lv Xianglin	Liu Chang
Liu Jiangang	Liu Haiying	Qi Chunyu	Guan Jianzhao	Guan Chengda
Sun Dongya	Lao Tianying	Du Longjiang	Li Na	Li Wenyang
Li Yunpeng	Yang Qingrui	Wu Leixiang	Yu Yang	Yu Xiao
Zhang Min	Zhang Jing	Zhang Zhihao	Zhang Xiaoming	Zhang Haitao
Zhang Haiping	Lu Qin	Chen He	Zhou Bo	Zheng Hao
Meng Yuan	Zhao Wei	Zhao Shilin	Zhao Jinyong	Hao Chunfeng
Hu Peng	Liu Changshun	Duan Xudong	Jiang Shan	Jiang Xiaoming
Luo Huihuang	Qin Wei	Jia Ling	Gao Jijun	Guo Chongshan
Tao Yuan	Huang Hai	Huang Xin	Huang Aiping	Chang Xiaomin
Qu Xiaodong	Ge Jinjin	Dong Fei	Dong Hao	Han Zhen
Xie Ying	Guan Xiaoyan	Ju Qianqian	Wei Zheng	Zhuge Yisi

Translators: Guo Chongshan Li Wenyang

Foreword

Rivers are crucial pathways in the global water cycle, and an important link for the transmission of energy, material and information flows in water ecosystems. Rivers are also the cradle of human civilization and the birthplace of the four ancient civilizations, having an important impact on social and economic development. The Night Light Index (NLI) is used to evaluate the social and economic development in a river basin, and almost all areas with a high NLI are situated along rivers, in the middle and lower reaches of river basins and in estuaries and bay areas. For example, the San Francisco Bay Area, the New York Bay Area, the Guangdong-Hong Kong-Macao Greater Bay Area and the Tokyo Bay Area, which are the most economically dynamic areas in the world, are all located in estuaries.

Rivers provide a variety of products and services for the development of human societies, including freshwater resources, aquatic products, hydropower generation, navigation, leisure and entertainment, and climate regulation. The freshwater ecosystems, composed of rivers, lakes and wetlands, are one of the most biodiverse regions on Earth, but the development of human societies has also affected rivers to varying degrees. *The World's Forgotten Fishes*, a report published in 2021 by 16 organizations including the World Wide Fund for Nature (WWF), states that populations of migratory freshwater fish have fallen by 76% and mega-fish by a catastrophic 94%. Ensuring that human societies and river ecosystems develop in a healthy manner and on an equal basis is an important issue for both scientific research and river conservation.

The United Nations 2030 Agenda for Sustainable Development emphasizes the role of water in supporting economic and social development and environmental protection amidst sustainable development, as well as the important role of healthy water ecosystems in sustainable development. To this end, the ecological value of rivers, the value of water infrastructure, the value of water supply, sanitation and hygiene services, the value of water for food and agriculture, and the value of water supply to energy, industry and commerce, and the cultural value of water should be fully recognized, measured and evaluated, and incorporated into national policymaking, which is vital to achieve sustainable water resources management and the United Nations 2030 Sustainable Development Goals (SDGs). Based on the concept of a "River of Happiness", China Institute of Water Resources and Hydropower Research (IWHR) has established the River Happiness Index that integrates five criteria, namely water security, water resources, water environment, water ecology and water culture, by drawing on as much as possible the advanced international experience in river protection, and selected 15 representative rivers around the world to scientifically evaluate the problems facing the rivers from a more comprehensive perspective and to understand the relationship between rivers and social development with a broader vision.

It is the first time globally to index the happiness of the world's rivers, hence an exploratory endeavor. The readers may find this report wanting in certain aspects, brought by the limitations in theory and experiences, and the insufficiency of basic data. Therefore, the research team wishes for our readers' understanding and advice so that we can improve. We look forward to joining hands with you to promote the world's rivers toward becoming Rivers of Happiness.

Contents

Chapter 1

River of Happiness: Definition and Connotation

Rivers are the blood vessels of the earth, the cradles of human civilizations, and the nexus of regional development in the river basins. The United Nations 2030 Sustainable Development Goals clearly stated that river ecosystems should be protected and restored, and water pollution should be reduced, so that everyone can have access to safe drinking water.

Based on the recognitions above, we envision a home Earth with all its waters, represented by rivers, being "happy" ones that can not only maintain a healthy ecosystem for themselves, but also serve to advance the well-being of the people along its blue and green, and bringing benefits and happiness to the whole society. Specifically, a river of happiness can have the following criteria: the prerequisite is its own health; a core function is to provide quality ecosystem products; an essential requirement is to sustain socio-economic development; an comprehensive manifestation is the harmony between man and water; and the fundamental gauge is whether it helps fulfill a sense of security and satisfaction for the people along the river.

Based on such understanding, the concept of **"River of Happiness"** is defined as such:

A River of Happiness is a river that maintains its own health, supports the sustainable socio-economic development of the basin and the region, enables harmony between human and water, and fulfills a high sense of security, gain and satisfaction to the people in the watershed. A River of Happiness is hence characterized by five dimensions: protecting people's life and safety, providing reliable water resources, creating a livable environment, sustaining a healthy aquatic ecosystem, and enhancing water-related culture and governance[1].

[1] IWHR's Research Team on "River of Happiness", Analysis of the connotation and index system for the River of Happiness, China Water Resources, 2020,23:1-4.

Fundamental guarantee: life and property safety protection

Throughout human history, flood has been the greatest long-term threat facing mankind. Looking back to the past, overflowing rivers have caused catastrophic destructions and brought huge damages to people's lives and properties, hence affecting social stability and economic development, and hindering the progress of social civilization and advancement. To better control and manage flood risks, is the way to safeguard people's well-being, sustain high-quality development, and bring **"sense of safety and security"** to the people along the rivers.

Basic function: high-quality and reliable water supply

Water is the source of life and is vital for human survival and social development. Rivers serve the society with quality and reliable water resources for people's daily use, for industries to run, and for agriculture to thrive. To provide **"safe and reliable"** water resources for a better life is the way that rivers fulfill their functions in terms of public service.

"Name card": Livable environment

The environment that rivers create is a key factor affecting human settlement and people's life quality, which are closely linked to surrounding environment. To ensure a livable environment along the riverside, it is necessary to protect and improve the water environment quality of both natural rivers and artificial water bodies in the urban and rural areas. To create an environment that is **"clean, beautiful and comfortable"** is the way to enable a more pleasant and happy life for the people.

Optimal state: healthy aquatic ecosystem

A healthy aquatic ecosystem is not only a key foundation for the sustainable development of human society, but also one of the most universal welfare that nature brings to people. The optimal state of a river of happiness is a river that can maintain the health of the aquatic systems, restore water environment, ensure the stability of the river ecosystem, and thus enable the **"harmony between man and nature"**.

Spiritual value: culture and governance enhancement

Culture, as the vein of a nation, gives the people a sense of belonging, and empowers a happy and satisfying life. In the long-term practice of water governance, all countries and regions in the world have created their own cultures about water, carrying the distinctive national characteristics and spirits. In the context of gearing up water civilization, to respect and protect rivers by regulating people's behaviors, correcting their wrong doings and developing water ethics has become the new standards for all mankind in face of river-related issues. To better protect, inherit and innovate the traditional water-related culture by raising public awareness and engagement in water governance is the way to make rivers better serve people as the **"spiritual homeland and soul of civilization"**.

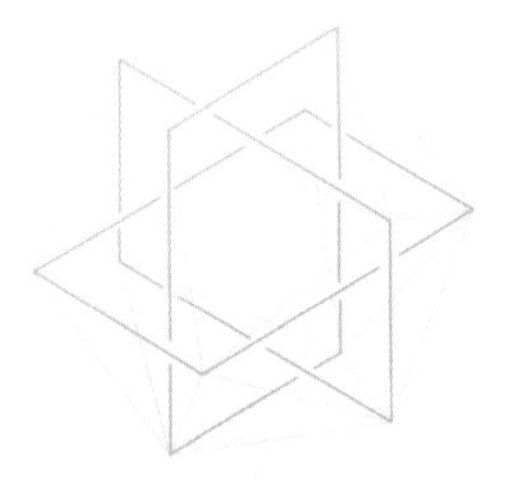

Chapter 2

River Happiness Index (RHI) and Its Evaluation Criteria

Evaluated through the five dimensions of a River of Happiness, abbreviated as safety, reliability, livability, health and culture, the River Happiness Index (RHI) sets up five primary indicators that comprise of Capacity of Life and Property Safety Protection (CSP), Reliability of Water Resources (RWR), Livability of Water Environment (LWE), Health of Aquatic Ecosystem (HAE), and Prosperity of Water-related Culture (PWC). There are 18 secondary indicators and certain tertiary indicators under the five primary indicators (Table 2.1).

Table 2.1 River Happiness Indexation

Primary indicators	Secondary indicators	Tertiary indicators
CSP(Capacity of Life and Property Safety Protection)	1. FMR (Flood-Induced Mortality Rate)	
	2. EIR (Economic Impact Rate)	
	3. PAC (Flood Disaster Prevention and Adaptation Capacity)	
RWR (Reliability of Water Resources)	4. AWP (Available Water Volume Per Capita)	
	5. WSR (Water Supply Reliability)	RIA (Rate of Actual Irrigated Areas)
	6. CSD (Capacity for Supporting High-Quality Development)	WUR (Water Resources Utilization Rate)
		GOW (GDP Output Per Cubic Meter of Water Use)
	7. LSI (Life Satisfaction Index)	GPC (GDP Per Capita)
		GINI (Gini Coefficient)
		ALE (Average Life Expectancy)
LWE (Livability of Water Environment)	8. PGW (Proportion of Water Bodies with Good Water Quality)	
	9. PSD (Percentage of Population with Safely Managed Drinking Water Services)	
	10. WTR (Urban Wastewater Treatment Rate)	
	11. WFI (Waterfront Index)	
HAE (Health of Aquatic Ecosystem)	12. VIH (Variation Index of Eco-Hydrological Process)	
	13. LCI (River Longitudinal Connectivity Index)	
	14. FEI (Fish Endangered Index)	
	15. STM (Sediment Transport Modulus)	
PWC (Prosperity of Water-related Culture)	16. CPI (Water Culture Protection and Inheritance Index)	
	17. MCI (Modern Water Culture Creation and Innovation Index)	
	18. PAG (Public Awareness and Involvement in Water Governance)	

The formulae for calculating the River Happiness Index are shown in (2.1)-(2.3).

$$\mathrm{RHI}=\sum_{i=1}^{5} F_i\, w_i^f \tag{2.1}$$

$$F_i=\sum_{j=1}^{J} S_{i,j}\, w_{i,j}^s \tag{2.2}$$

$$S_{i,j}=\sum_{k=1}^{K} T_{i,j,k}\, w_{i,j,k}^t \tag{2.3}$$

Where: RHI is the River Happiness Index; F_i is the score of the ith primary indicator, where i is the subscript of the primary indicator with an assigned value from 1 to 5 that corresponds to CSP, RWR, LWE, HAE, and PWC, respectively; w_i^f is the weight of the ith primary indicator; $S_{i,j}$ is the score of the jth secondary indicator under the ith primary indicator, where j is the subscript of the secondary indicator with an assigned value from 1 to J; $w_{i,j}^s$ is the weight of the jth secondary indicator under the ith primary indicator; $T_{i,j,k}$ is the score of the kth tertiary indicator under the jth secondary indicator under the ith primary indicator, where k is the subscript of the tertiary indicator with an assigned value from 1 to K; $w_{i,j,k}^t$ is the weight of the kth tertiary indicator under the jth secondary indicator under the ith primary indicator.

The River Happiness Index evaluation is based on a 100-point scoring system. A river is determined as a River of Happiness when its RHI scores 85 points or more (Table 2.2). All the indicators are graded (Table 2.3) as "good" when it scores 85 points or more. The connotations of all the indicators and their evaluation methods and standards can be found in the Appendix.

Table 2.2 Grading of the River Happiness Index

RHI	Grade		
RHI ⩾ 95	very happy		
85 ⩽ RHI < 95	happy		
60 ⩽ RHI < 85	medium	80 ⩽ RHI < 85	medium high
		70 ⩽ RHI < 80	medium
		60 ⩽ RHI < 70	medium low
RHI < 60	unhappy		

Table 2.3 Grading of the indicators under the River Happiness Index

Indicator scoring V[①]	Grade		
$V \geqslant 95$	excellent		
$85 \leqslant V < 95$	good		
$60 \leqslant V < 85$	medium	$80 \leqslant V < 85$	medium high
		$70 \leqslant V < 80$	medium
		$60 \leqslant V < 70$	medium low
$V < 60$	poor	$30 \leqslant V < 60$	quite poor
		$V < 30$	extremely poor

① V indicates F_i, $S_{i,j}$, or $T_{i,j,k}$.

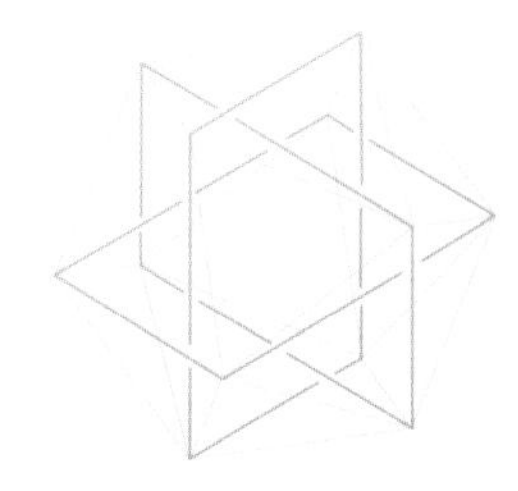

Chapter 3

Evaluation Scope of the World's Rivers

This report selects 15 representative rivers worldwide, in consideration of continental representation, data availability, and the history and the location significance of the rivers. They are the Amazon River, the Colorado River, the Congo River, the Danube River, the Euphrates River (a short name for the Euphrates-Tigris[1]), the Ganges River, the Mississippi River, the Murray-Darling River, the Nile River, the Rhine River, the Saint Lawrence River, the Thames River, the Volga River, the Yangtze River, and the Yellow River.

In terms of their spatial distribution, the 15 evaluated rivers include four in Asia, four in Europe, three in North America, one in South America, two in Africa, and one in Oceania. (Table 3.1).

Table 3.1 Basic information of the 15 rivers selected for RHI evaluation

River	Continent	Amounts of riverine countries	Length/km	Drainage area/10,000 km^2
Amazon River	South America	8	6,448	596.8
Colorado River	North America	2	2,330	65.3
Congo River	Africa	10	4,700	368.9
Danube River	Europe	10	2,857	81.7
Euphrates River	Eurasia	6	2,800	107.0
Ganges River	Asia	4	2,527	176.4
Mississippi River	North America	2	6,021	322.0
Murray-Darling River	Oceania	1	3,672	155.7
Nile River	Africa	11	6,670	335.0

[1] In this report, the Euphrates River for the Euphrates is used as the short name for the Euphrates and the Tigris rivers. The length of the longer one is used to represent the length of the Euphrates, while the drainage areas and populations of both river basins are summed up as the total area and population of the Euphrates.

Continued

River	Continent	Amounts of riverine countries	Length/km	Drainage area/10,000 km^2
Rhine River	Europe	9	1,320	16.1
Saint Lawrence River	North America	2	3,058	105.4
Thames River	Europe	1	346	1.3
Volga River	Europe	1	3,645	138.0
Yangtze River	Asia	1	6,397	178.4
Yellow River	Asia	1	5,464	79.5

Most of the rivers evaluated in this report are transboundary ones. For the Congo River, the Danube River and the Nile River, each of them flows through 10 countries, more than any of the other 12 rivers; the Rhine River and the Amazon River flow through 9 and 8 countries, respectively. Only the Murray-Darling River, the Volga River, the Thames River, the Yangtze River and the Yellow River flow within one country.

The total length of the rivers evaluated is over 60,000 km, including some of the world's longest rivers such as the Nile River (6,67 km), the Amazon River (6,448 km), and the Yangtze River (6,397 km).

The evaluation also covers the birthplaces of the four ancient civilizations: the Nile River, the Euphrates-Tigris River, the Ganges River and the Yellow River basins. The total catchment area of the rivers amounted to 26.9 million km^2, about 18% of the total global land area, with the world top three largest river basins being included, which are the Amazon River Basin (5.968 million km^2), the Congo River Basin (3.7 million km^2), and the Mississippi River Basin (3.22 million km^2).

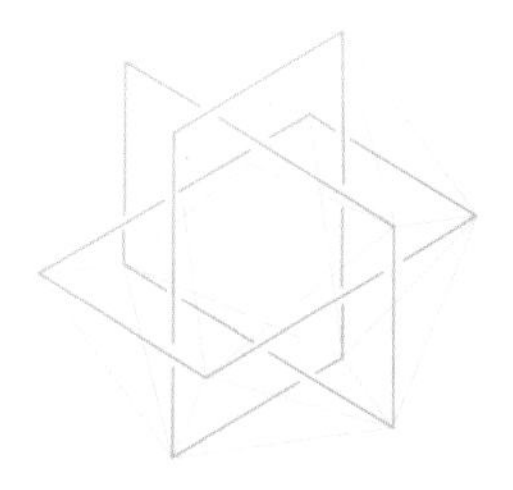

Chapter 4

Findings

Amazon River 71.5 points

Colorado River 81.8 points

Congo River 70.1 points

Danube River 79.7 points

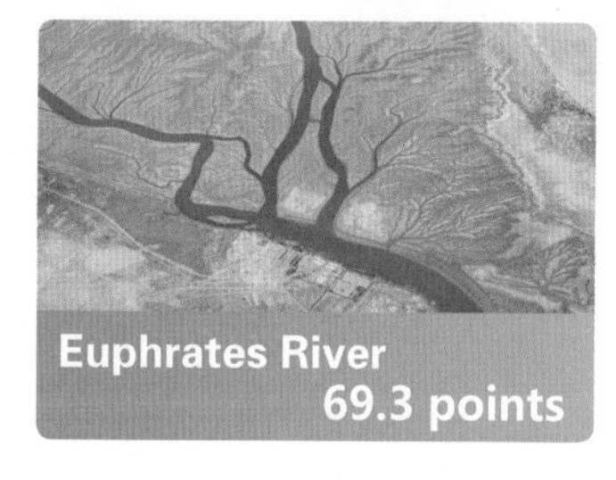
Euphrates River 69.3 points

Ganges River 65.6 points

Mississippi River 80.1 points

Murray–Darling River 75.6 points

Nile River 62.1 points

Rhine River 86.6 points

Saint Lawrence River 84.6 points

Thames River 81.9 points

Volga River 79.0 points

Yangtze River 80.8 points

Yellow River 78.8 points

4.1 Amazon River

The Amazon River, with a length of 6,448 km, is the world's second longest river and the largest river by discharge volume (2.19 million m^3/s) and by drainage area (5.968 million km^2). Located at the northern part of South America and originated from the middle section of the Andes Mountains in Peru, the Amazon flows from west to east with many tributaries along its way, forming a huge river network. The Amazon flows across a number of countries including Peru, Ecuador, Colombia, Venezuela, Guyana, Suriname, Bolivia, Brazil and eventually joins into the Atlantic Ocean near the island of Marajó in Brazil.

Being crowned as "the lungs of the earth", the Amazon River boasts vastly rich ecological resources in its watershed. It is home to over 20,000 plant species, nearly 3,000 fish species and over 1,600 bird species, of which the typical ones include Caimans, freshwater turtles, and aquatic mammals such as manatees and freshwater dolphins, and terrestrial animals like jaguar, jaguarundi, tapir, capybara, and armadillo.

The RHI of the Amazon River receives an overall score of 71.5 points, and is graded as "medium" (Figure 4.1).

Capacity of Life and Property Safety Protection (CSP). The CSP of the Amazon River receives a score of 66.3 points, which is graded as "medium low". For the three secondary indicators under this primary indicator, the Flood-Induced Mortality Rate (FMR) and the Economic Impact Rate (EIR) both score

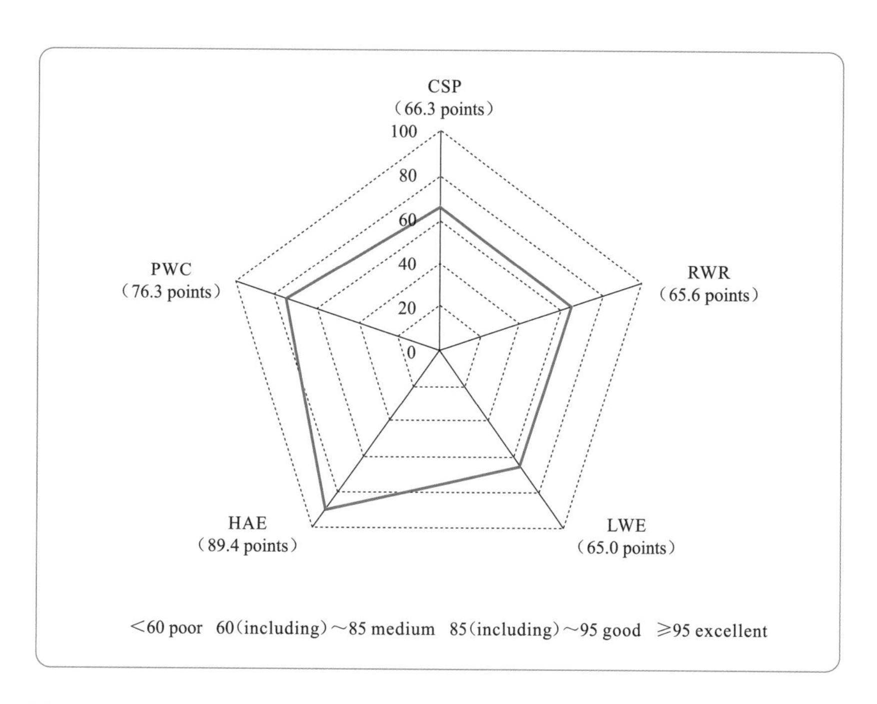

Figure 4.1 Scores of five primary indicators of RHI of the Amazon River

60.0 points and are graded as "medium low", while the Flood Disaster Prevention and Adaptation Capacity (PAC) scores 75.8 points and is graded as "medium".

Reliability of Water Resources (RWR). The RWR of the Amazon River receives a score of 65.6 points, which is graded as "medium low". For the four secondary indicators under the RWR, the Available Water Volume Per Capita (AWP), with an amount of 208,014 m^3 for the Amazon River, scores 100 points and is graded as "excellent"; the Water Supply Reliability (WSR) scores 60.1 points and is graded as "medium low"; the Capacity for Supporting High-Quality Development (CSD) and the Life Satisfaction Index (LSI) both receive scores between 50-60 points and thus are graded as "quite poor".

Livability of Water Environment (LWE). The LWE of the Amazon River receives a score of 65.0 points and is graded as "medium low". Among the four secondary indicators, the Proportion of Water Bodies with Good Water Quality (PGW) and the Percentage of Population with Safely Managed Drinking Water Services (PSD) score 71.0 points and 74.8 points respectively and are both graded as "medium"; the Urban Wastewater Treatment Rate (WTR) scores 41.5 points, which falls into the grade of "quite poor"; and the Waterfront Index (WFI) scores 64.9 points and is graded as "medium low".

Health of Aquatic Ecosystem (HAE). The HAE of the Amazon River scores 89.4 points and is graded as "good". And for the four secondary indicators under the HAE, the Variation Index of Eco-Hydrological Process (VIH) and the Sediment Transport Modulus (STM) score 81.5 points and 83.4 points, and are both graded as "medium high"; while the River Longitudinal Connectivity Index (LCI) and the Fish Endangered Index (FEI) both score above 95 points and are thus graded as "excellent".

Prosperity of Water-related Culture (PWC). The PWC of the Amazon River scores 76.3 points and is graded as "medium". For the three secondary indicators, the Water Culture Protection and Inheritance Index (CPI) scores 91.9 points and reaches the grade of "good"; and the Modern Water Culture Creation and Innovation Index (MCI) and the Public Awareness and Involvement in Water Governance (PAG) score 66.1 points and 65.8 points respectively, and are graded as "medium low".

The evaluation results of the RHI of the Amazon River show that: first, water disasters such as flood have exerted great impacts on the socio-economic development in the river basin; second, despite its abundance, the water resources of the Amazon River is still far from sufficient to support the high-quality development of the countries and the regions along the river; third, the living environment of the riparian areas is not quite good, neither are the proportion of good-quality water resources, the harmony of cities and riverfront areas, and the sewage treatment; fourth, the basin has been doing well in the protection and inheritance of water heritages, but not so much in the innovation of those heritages and creation of modern culture, which need to be further enhanced to enrich the spiritual life of the people in the basin.

4.2 Colorado River

The Colorado River, starting in the central Rocky Mountains of Colorado in western U.S., is sourced from vast areas of melting snowpack from the Mountains and flows southwest with a great part draining into the Gulf of California, while a small part running southward into the Salton Sea in Mexico. The 2,330 km long river drains an expansive, arid watershed of 653,000 km^2 that encompasses several states in the U.S. and Mexico.

The hydroelectricity from the Colorado River is a key supplier of peaking power on the southwest electric grid of the U.S. often called "America's Nile", the Colorado is so intensively managed that each drop of its water is used an average of 17 times in a single year. The over-exploitation dewatered the lower Colorado, and flowing water can no longer be seen in the estuary since 1993. Therefore, the Colorado River became one of the rivers that have the earliest legislations on its development and management.

The RHI of the Colorado River receives an overall score of 81.8 points, and is graded as "medium-high" (Figure 4.2).

Capacity of Life and Property Safety Protection (CSP). The CSP of the Colorado River scores 90.9 points, which is graded as "good". For the three secondary indicators, the Flood-Induced Mortality Rate (FMR) and the Economic Impact Rate (EIR) both score 90.0 points and are graded as "good", while the Flood Disaster Prevention and Adaptation Capacity (PAC) scores 92.3 points and is graded as "good".

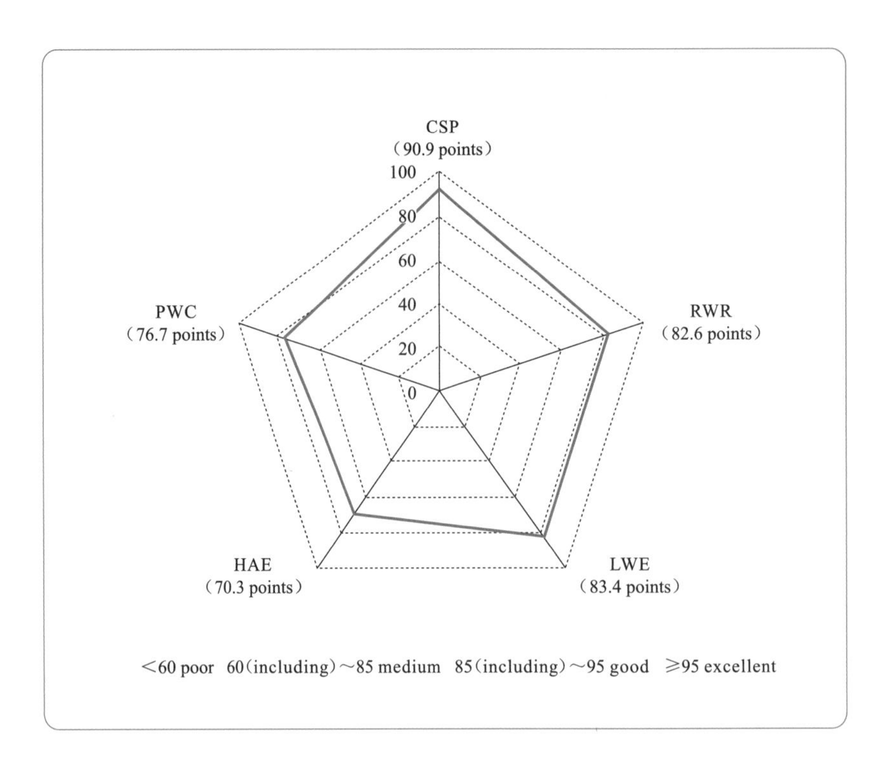

Figure 4.2 Scores of five primary indicators of RHI of the Colorado River

Reliability of Water Resources (RWR). The RWR of the Colorado River receives a score of 82.6 points, which is graded as "medium high". For the four secondary indicators under the RWR, the Available Water Volume Per Capita (AWP), with an amount of 6,981 m^3 for the Colorado River, scores 92.7 points and is graded as "good"; the Water Supply Reliability (WSR) and the Capacity for Supporting High-Quality Development (CSD) score 80.2 points and 80.6 points respectively and are both graded as "medium high", and the Life Satisfaction Index (LSI) receives a score of 79.6 points and is graded as "medium".

Livability of Water Environment (LWE). The LWE of the Colorado River receives a score of 83.4 points and is graded as "medium high". Among the four secondary indicators, the Proportion of Water Bodies with Good Water Quality (PGW) scores 78.8 points and is graded as "medium"; the Percentage of Population with Safely Managed Drinking Water Services (PSD) scores 96.3 points and reaches the grade of "excellent"; the Urban Wastewater Treatment Rate (WTR) scores 92.2 points, which is graded as "good"; and the Waterfront Index (WFI) scores 62.1 points and is graded as "medium low".

Health of Aquatic Ecosystem (HAE). The HAE of the Colorado River scores 70.3 points and is graded as "medium". For the four secondary indicators, the Variation Index of Eco-Hydrological Process (VIH) scores 37.4 points and is graded as "quite poor"; the River Longitudinal Connectivity Index (LCI) and the Fish Endangered Index (FEI) both score between 70-80 points and are thus graded as "medium"; and the Sediment Transport Modulus (STM) scores 100.0 points and is graded as "excellent".

Prosperity of Water-related Culture (PWC). The PWC of the Colorado River scores 76.7 points and is graded as "medium". For the three secondary indicators, the Water Culture Protection and Inheritance Index (CPI) scores 66.0 points and is graded as "medium low"; the Modern Water Culture Creation and Innovation Index (MCI) scores 82.3 points and is graded as "medium high"; and the Public Awareness and Involvement in Water Governance (PAG) scores above 85.0 points and reaches the grade of "good".

The evaluation results of the RHI of the Colorado show that: first, despite its rich water resources, further efforts are needed to improve the capacities in supplying water resources and supporting regional development; second, the riverfront environment is moderate in its livability, with the proportion of good-quality water bodies and the waterfront index are relatively low; third, the health status of the aquatic system in the basin is not optimistic, as the variation of eco-hydrological process and the longitudinal connectivity become noticeable problems; fourth, water-related culture and governance are not satisfactory so far, the protection and inheritance of which should be specifically enhanced.

4.3 Congo River

The Congo River, the deepest river in the world, rises in the highlands of northern Zambia and flows across the famous Congo basin in equatorial Africa, with a typical basin shape. It is situated between the Sahara to the north, the Atlantic Ocean to the south and west, and the region of the East African lakes to the east, with a length of 4,700 km, a drainage basin covering an area of about 3.7 million km^2, a width of several kilometers, a depth of 100 m to 200 m, an average annual runoff of about 41,000 m^3/s at its mouth, and an annual inflow of 1.3 trillion m^3 into the Atlantic Ocean. The main stream and tributaries of the Congo River stretch into Zambia, Tanzania, Angola, Central Africa, Cameroon, the Republic of the Congo and the Democratic Republic of the Congo before draining into the Atlantic Ocean. With its many tributaries, the Congo River forms a dense network of rivers, with its main stream running along the edge of the Congo River basin to take the form of a large arc protruding northward and cross the Equator twice, and most of its tributaries flowing out of equatorial zones where heavy rainfall occurs.

Due to the rapids and waterfalls scattered over its main stream and tributaries, the Congo River boasts abundant hydropower resources with theoretical hydro energy reserves of 390 GW, ranking first among the major rivers in the world. The installed capacity of exploitable hydro energy resources is approximately 156 GW. The dense tributaries and sub-tributaries in the river basin form the Africa's largest network of navigable waterways, which is the continent's most important navigation system, with a navigable length of around 20,000 km across the main stream and tributaries. The Congo river basin is home to the second largest tropical rainforest in the world, after the Amazon rainforest in South America, and rich biological resources, including nearly 700 known fish species, of which 80% are endemic.

The RHI of the Congo River gets an overall score of 70.1 points, and is graded as "medium" (Figure 4.3).

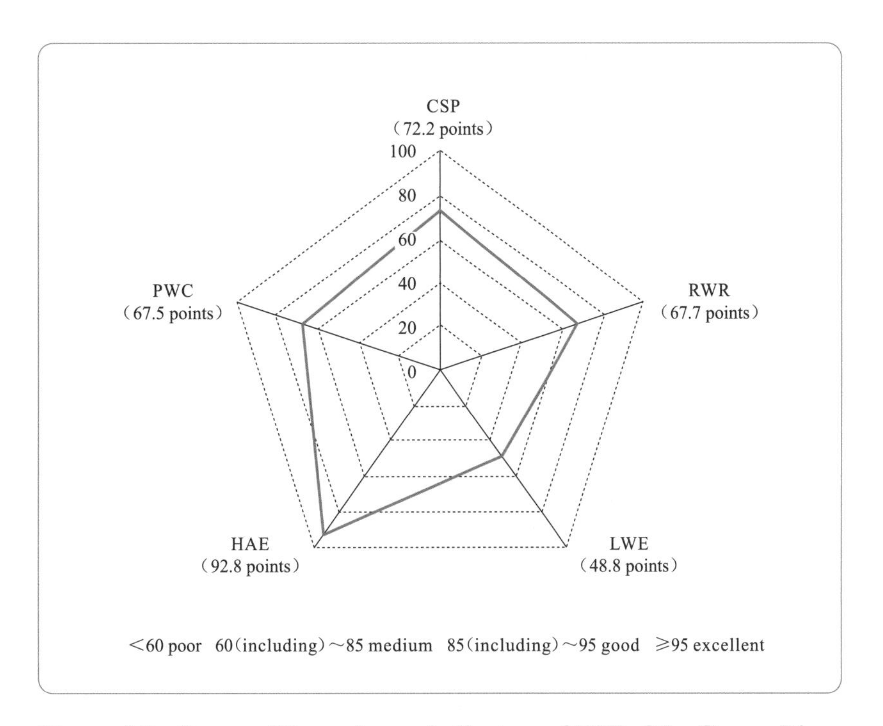

Figure 4.3 Scores of five primary indicators of RHI of the Congo River

Capacity of Life and Property Safety Protection (CSP). The CSP of the Congo River scores 72.2 points, which is graded as "medium". For the three secondary indicators, the Flood-Induced Mortality Rate (FMR) scores 75.0 points and is graded as "medium"; the Economic Impact Rate (EIR) scores 80.0 points and is graded as "medium high", while the Flood Disaster Prevention and Adaptation Capacity (PAC) scores 64.3 points and is thus graded as "medium low".

Reliability of Water Resources (RWR). The RWR of the Congo River receives a score of 67.7 points, which is graded as "medium low". For the secondary indicators, the Available Water Volume Per Capita (AWP) scores 100.0 points and is graded as "excellent"; the Water Supply Reliability (WSR) scores 60.1 points and is graded as "medium low"; the Capacity for Supporting High-Quality Development (CSD) scores 78.7 points and is graded as "medium", and the Life Satisfaction Index (LSI) receives a score of 39.8 points and is graded as "quite poor".

Livability of Water Environment (LWE). The LWE of the Congo River receives a score of 48.8 points and is graded as "quite poor". Among the four secondary indicators, the Proportion of

Water Bodies with Good Water Quality (PGW) scores 63.7 points and is graded as “medium low”; the Percentage of Population with Safely Managed Drinking Water Services (PSD) scores 41.4 points and falls into the grade of “quite poor”; the Urban Wastewater Treatment Rate (WTR) scores only 1.5 points, which is graded as “extremely poor”; and the Waterfront Index (WFI) scores 85.0 points and is graded as “good”.

Health of Aquatic Ecosystem (HAE). The HAE of the Congo River scores 92.8 points and is graded as “good”. For the secondary indicators, the Variation Index of Eco-Hydrological Process (VIH) scores 90.3 points and is graded as “good”; the River Longitudinal Connectivity Index (LCI) scores 99.6 points and is graded as “excellent”; the Fish Endangered Index (FEI) and the Sediment Transport Modulus (STM) score 94.7 points and 87.4 points respectively and are both graded as “good”.

Prosperity of Water-Related Culture (PWC). The PWC of the Congo scores 67.5 points and is graded as “medium low”. For the secondary indicators, the Water Culture Protection and Inheritance Index (CPI) scores 73.4 points and is graded as “medium”; the Modern Water Culture Creation and Innovation Index (MCI) and the Public Awareness and Involvement in Water Governance (PAG) score 61.4 points and 65.7 points respectively and are graded as “medium low”.

The evaluation results of the RHI of the Congo show that: first, the inefficiency of urban wastewater treatment and limited coverage of safe drinking water service are the most protruding problems that need to be urgently dealt with; second, the flood disaster prevention and adaptation capacity is low, which is the weakness of the Congo River and needs to be further improved; third, both the per capita GDP and the people’s happiness levels are low, which is the key issue in water supply reliability; fourth, there is still much progress to be made in the creation and innovation of modern water culture, and the function of water-related economy and culture in driving social development and improving people’s quality of life needs to be further enhanced.

4.4 Danube River

The Danube River, the second largest river in Europe, flows for 2,857 km and drains a basin of 817,000 km^2, with an average annual discharge of 6,430 m^3/s and an average annual runoff of 203 billion m^3 at its mouth. Originating in southwestern Germany, the Danube has a dense network of tributaries and flows from west to east through countries such as Austria, Slovakia, Hungary, Croatia, Serbia, Bulgaria, Romania, Moldova and Ukraine before emptying into the Black Sea. It is one of the rivers whose main stream runs through the most countries in the world.

The Danube River basin supports 83 million people in Europe and provides drinking water for 20 million people; it is also an international river of great navigational value and is connected to the Rhine River system through a canal to form a water transportation network in Central Europe. The Danube is rich in hydropower resources, with theoretical reserves of 50 billion kWh; there are several hydropower stations built on the main stream. Historically, the Danube was a unique habitat for 10 rare fish species in the world, with 103 precious fish species and 88 freshwater molluskc species. The inland navigation projects, and over-exploitation of natural resources and building of dams and dykes on the Danube in the 1970s and 1980s had greatly damaged the ecological environment of the river, resulting in the shrinkage of swamps and wetlands, the disappearance of numerous flood plains and flood detention areas, serious water pollution and severe degradation of the water ecology there. In recent years, the joint efforts of many countries to improve the ecological environment of the Danube River have produced remarkable results, with the water quality greatly improved and the number of aquatic species doubling the level in the 1980s.

The RHI of the Danube River gets an overall score of 79.7 points, and is graded as "medium" (Figure 4.4).

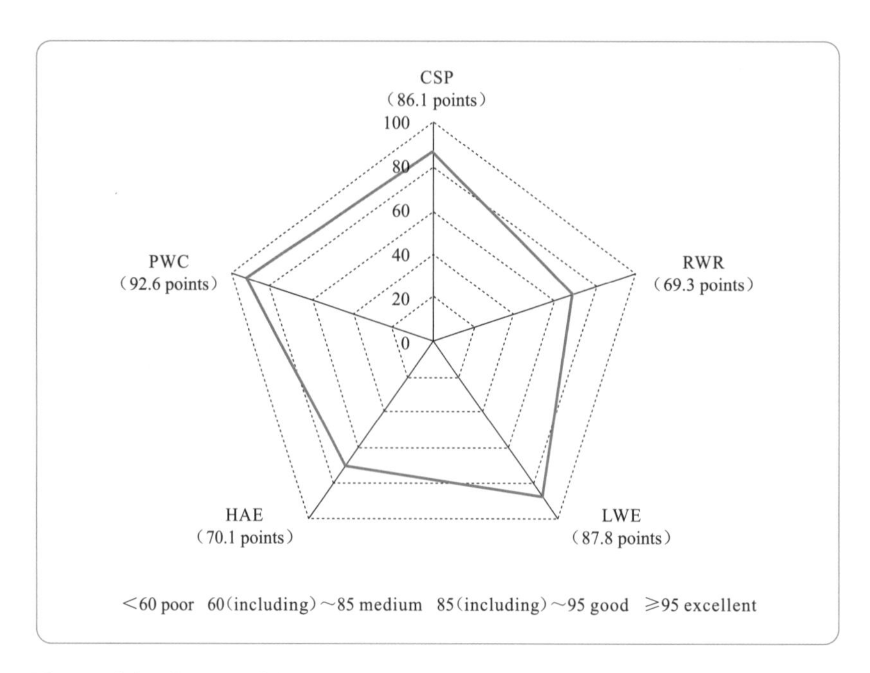

Figure 4.4 Scores of five primary indicators of RHI of the Danube River

Capacity of Life and Property Safety Protection (CSP). The CSP of the Danube River scores 86.1 points, which is graded as "good". For the three secondary indicators, the Flood-Induced Mortality Rate (FMR), which is 0.49 per million people in the basin, scores 90.0 points and is graded as "good"; the Economic Impact Rate (EIR), with a number of 0.37%, scores 85.0 points and is also graded as "good", and the Flood Disaster Prevention and Adaptation Capacity (PAC) scores 84.1 points and is graded as "medium high".

Reliability of Water Resources (RWR). The RWR of the Danube River receives a score of 69.3 points, which is graded as "medium low". For the secondary indicators, the Available Water Volume Per Capita (AWP), with its amount of 2933.3 m^3, scores 83.0 points and is graded as "medium high"; the Water Supply Reliability (WSR) scores 18.7 points and is graded as "extremely poor"; the Capacity for Supporting High-Quality Development (CSD) scores 100.0 points and is graded as "excellent", and the Life Satisfaction Index (LSI) scores 88.2 points and is graded as "good".

Livability of Water Environment (LWE). The LWE of the Danube River scores 87.8 points and is graded as "good". Among the secondary indicators, the Proportion of Water Bodies with Good Water

Quality (PGW) scores 98.0 points and is graded as "excellent"; the Percentage of Population with Safely Managed Drinking Water Services (PSD), which is 68.2%, scores 88.2 points and is graded as "good"; the Urban Wastewater Treatment Rate (WTR), which is 50%, scores 84.5 points and is graded as "medium high"; and the Waterfront Index (WFI) scores 75.1 points and is graded as "medium".

Health of Aquatic Ecosystem (HAE). The HAE of the Danube River scores 70.1 points and is graded as "medium". For the secondary indicators, the Variation Index of Eco-Hydrological Process (VIH) scores 55.6 points and is graded as "quite poor"; the River Longitudinal Connectivity Index (LCI) scores 72.6 points and is graded as "medium"; the Fish Endangered Index (FEI) scores 81.3 points and is graded as "medium high"; and the Sediment Transport Modulus (STM), which is 18.8 $t/(km^2 \cdot a)$ for the river, scores 75.8 points and is graded as "medium".

Prosperity of Water-related Culture (PWC). The PWC of the Danube River receives a score of 92.6 points and is graded as "good". For the secondary indicators, the Water Culture Protection and Inheritance Index (CPI) scores 97.6 points and is graded as "excellent"; the Modern Water Culture Creation and Innovation Index (MCI) and the Public Awareness and Involvement in Water Governance (PAG) score 88.4 points and 90.3 points respectively and are both graded as "good".

The evaluation results of the RHI of the Danube River show that: first, the Flood Disaster Prevention and Adaptation Capacity (PAC) is relatively low compared with another two secondary indicators under CSP to guarantee life and property safety; second, the Water Supply Reliability (WSR), which is graded as "extremely poor", is a key limiting factor to ensure the sustainability of water supply; third, the improvement of urban wastewater treatment and waterfront ; fourth, water-related culture and governance are not satisfactory so far, for which the protection and inheritance of water heritages should be specifically paid attention to.

4.5 Euphrates River

The Euphrates River, short for the Euphrates-Tigris River system, the largest in Southwest Asia, consists of the Tigris and the Euphrates, which rise in the mountains of eastern Turkey and flow southeast across northern Syria and through Iraq to the Persian Gulf. The Euphrates, as the longest river in West Asia, rises in the mountains of the Anatolian Plateau and the Armenian Highland in Turkey and flows through Turkey, Syria and Iraq, with a total length of 2,800 km, a drainage area of 690,000 km^2 and an average annual runoff of 37 billion m^3; the Tigris originates in the southern foot of the eastern Taurus Mountains southeast of the Anatolian Plateau in Turkey, and flows southeast through Turkey to form the border between Turkey and Syria and then goes directly into Iraq, with a total length of 1,950 km, a drainage area of 380,000 km^2 and an average annual runoff of about 40 billion m^3. The rivers join at Al-Qurnah to form the Shatt al-Arab before emptying into the Persian Gulf.

Having risen in close proximity, the Tigris and Euphrates diverge sharply in their upper courses, to a maximum distance of some 402 km. Their middle courses gradually approach each other, bounding a triangle of mainly barren limestone desert known as Al-Jazirah (Arabic: "The Island"). There the rivers have cut deep and permanent beds in the rock, so that their courses have undergone only minor changes since prehistoric times. The Mesopotamian plain nourished by the rivers was once the seat of ancient Babylonia, which gave birth to the Mesopotamian civilization, the oldest civilization in the world.

The RHI of the Euphrates River gets an overall score of 69.3 points, and is graded as "medium low" (Figure 4.5).

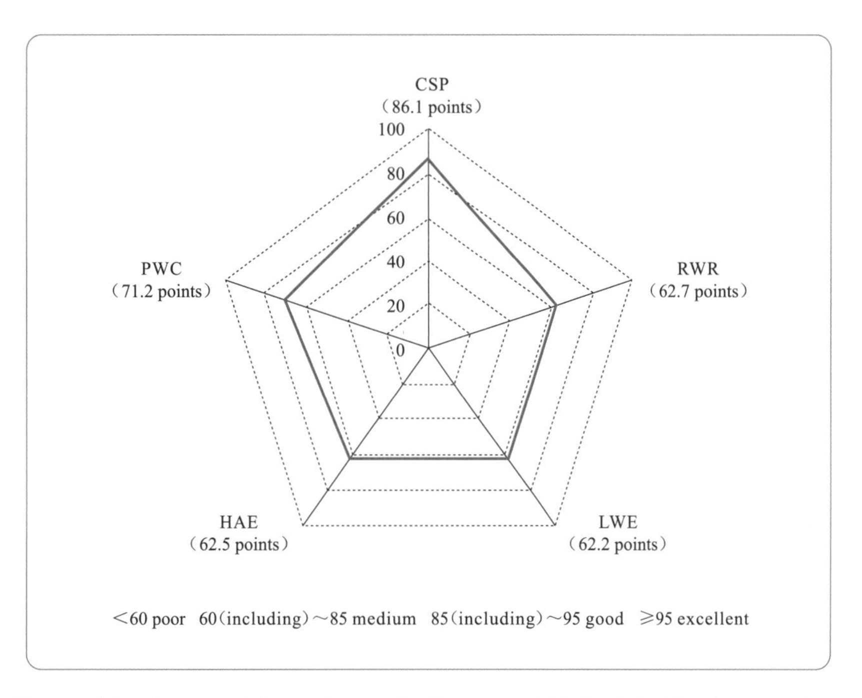

Figure 4.5 Scores of five primary indicators of RHI of the Euphrates River

Capacity of Life and Property Safety Protection (CSP). The CSP of the Euphrates River reaches "good" with a score of 86.1 points. Among the three secondary indicators under this primary indicator, the Flood-Induced Mortality Rate (FMR) and the Economic Impact Rate (EIR) score 90.0 points and 95.0 points, reaching "good" and "excellent", respectively, while the Flood Disaster Prevention and Adaptation Capacity (PAC) scores 76.6 points and is graded as "medium".

Reliability of Water Resources (RWR). The RWR of the Euphrates River receives a score of 62.7 points, which is graded as "medium low". For the four secondary indicators under the RWR, the Available Water Volume Per Capita (AWP) and the Water Supply Reliability (WSR) score 75.4 points and 79.0 points respectively and both reach "medium"; the Life Satisfaction Index (LSI) and the Capacity for Supporting High-Quality Development (CSD) score 59.2 points and 36.5 points respectively, and are both graded as "quite poor".

Livability of Water Environment (LWE). The LWE of the Euphrates River receives a score of 62.2 points and is graded as "medium low". Regarding the four secondary indicators, the Waterfront Index (WFI) gains 79.3 points and is graded as "medium"; the Percentage of Population with Safely Managed Drinking Water Services (PSD) and the Proportion of Water Bodies with Good Water Quality (PGW) score 66.2 points and 60.5 points and are both graded as "medium low"; and the Urban Wastewater Treatment Rate

(WTR) only scores 41.9 points, thus being graded as “quite poor”.

Health of Aquatic Ecosystem (HAE). The HAE of the Euphrates River scores 62.5 point and is graded as “medium low”. For the four secondary indicators under the HAE, the River Longitudinal Connectivity Index (LCI) obtains a score of 72.4 points and is graded as “medium”; the Sediment Transport Modulus (STM) and the Fish Endangered Index (FEI) both reach a “good” with scores of 93.7 points and 89.3 points respectively; while the Variation Index of Eco-Hydrological Process (VIH) scores 10.4 points and is graded as “extremely poor”.

Prosperity of Water-related Culture (PWC). The PWC of the Euphrates River receives 71.2 points and is graded as “medium”. Among the three secondary indicators, the Water Culture Protection and Inheritance Index (CPI) scores 83.3 points and is graded as “medium high”; the Modern Water Culture Creation and Innovation Index (MCI) and the Public Awareness and Involvement in Water Governance (PAG) score 60.0 points and 66.4 points respectively, and are both graded as “medium low”.

Based on the RHI, the evaluation results of the Euphrates River show that: first, the capacity of water resources is inadequate in supporting high-quality development, mainly posing a potential risk to the reliability of water resources in the Euphrates River basin; second, the social and economic development is hindered by the poor performance in urban wastewater treatment; third, the rather low point of the variation index of eco-hydrological process might become a major issue for bettering the living environment nearby the basin; fourth, the shortcoming of innovation and creation of modern water culture needs to be addressed for water culture protection.

4.6 Ganges River

Located in southern Asia and originated from the southern Great Himalayas, the Ganges River is formed when the Alaknanda and Bhagirathi rivers unite at Devaprayag, and eventually flows into the Bay of Bengal. The main stream of the Ganges has a total length of 2,527 km, and drains an area of 1.764 million km^2 across India, Nepal, China and Bangladesh.

The Ganges River, as the largest river in India, is one of the birthplaces of Indian civilization, and is not only a river held sacred by Hindus, but also the place where Buddhism began. The Ganges River is the most populous river basin in the world, which is inhabited by over 400 million people. The population density there is more than 390 per km^2. The Ganges River is witnessing serious water pollution, which has a major impact on both people and animals living in the basin. A 2021 report by the Central Pollution Control Board (CPCB) shows that in Bhagalpur, Bihar in the middle and lower reaches of the Ganges River, the concentration of fecal coliform in the reaches monitored by the water quality monitoring station was more than 64 times higher than the standard value set by the Indian government. Please note that only the reaches in India is assessed in this report.

The RHI of the Ganges River receives an overall score of 65.6 points and is graded as "medium low" (Figure 4.6).

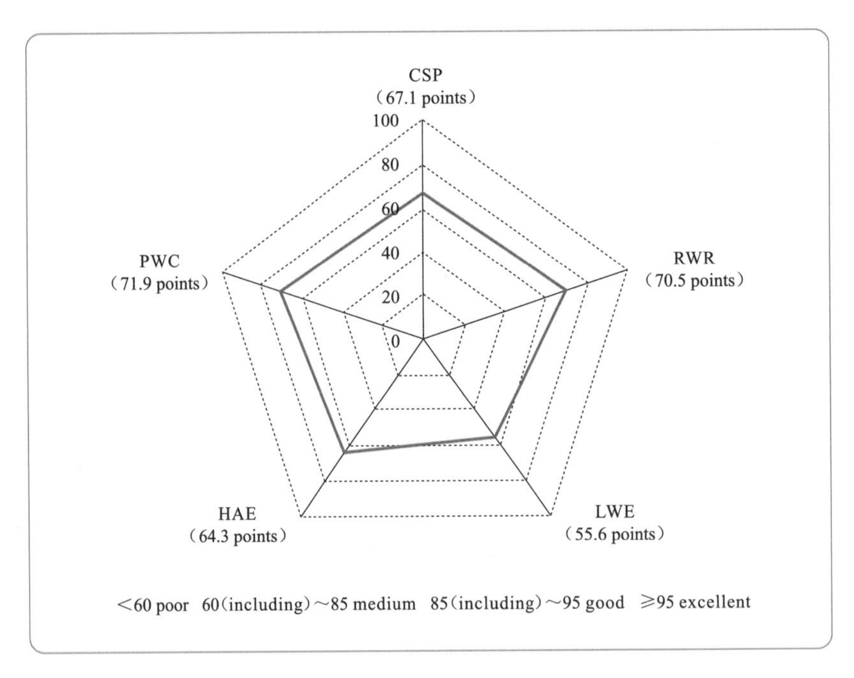

Figure 4.6 Scores of five primary indicators of RHI of the Ganges River

Capacity of Life and Property Safety Protection (CSP). The CSP of the Ganges River receives a score of 67.1 points and is graded as "medium low". For the three secondary indicators, the Flood Disaster Prevention and Adaptation Capacity (PAC) scores 74.1, which is graded as "medium"; the Economic Impact Rate (EIR) and the Flood-Induced Mortality Rate (FMR) are both graded as "medium low" with respective scores of 65.0 points and 60.0 points.

Reliability of Water Resources (RWR). The RWR of the Ganges River receives a score of 70.5 points, which is graded as "medium". For the four secondary indicators under the RWR, the Available Water Volume Per Capita (AWP) and the Water Supply Reliability score 88.2 points and 89.5 points, both reach the grade of "good"; while the Life Satisfaction Index (LSI) and the Capacity for Supporting High-Quality Development (CSD) are both graded as "quite poor" by scoring 52.9 points and 51.0 points respectively.

Livability of Water Environment (LWE). The LWE of the Ganges River gets a score of 55.6 points and is graded as "quite poor". Among the four secondary indicators, the Proportion of Water Bodies with Good Water Quality (PGW), the Percentage of Population with Safely Managed Drinking Water Services (PSD) and the Urban Wastewater Treatment Rate (WTR) are all graded as "quite poor" with scores of 50.6 points, 54.5 points and 32.1 points individually; while the Waterfront Index (WFI) reaches "good" through a score of 88.2 points.

Health of Aquatic Ecosystem (HAE). The HAE of the Ganges River scores 64.3 points and is graded as "medium low". For the four secondary indicators, the River Longitudinal Connectivity Index (LCI)

receives a score of 91.3 points and is graded as "good"; the Sediment Transport Modulus (STM) scores 83.7 points and is graded as "medium high"; the Fish Endangered Index (FEI) scores 70.0 points and is at the grade of "medium"; and the Variation Index of Eco-Hydrological Process (VIH) scores 21.7 points and is thus graded as "extremely poor".

Prosperity of Water-related Culture (PWC). The PWC of the Ganges River scores 71.9 points and reaches the grade of "medium". For the three secondary indicators, the Water Culture Protection and Inheritance Index (CPI) gains a score of 84.1 points and is graded as "medium high"; while the Modern Culture Creation and Innovation Index (MCI) and the Public Awareness and Involvement in Water Governance (PAG) are both at "medium low" by getting scores of 62.2 points and 65.4 points respectively.

The evaluation results from the RHI of the Ganges River reveal that: first, its water quality in general is quite poor, neither are the proportion of population who is accessible to safe drinking water and the urban wastewater treatment; second, the variation of eco-hydrological process is worrying, and might challenge the quality and stability of the Ganges ecosystem; third, the insufficient capacity of recovery from flood has weakened the river's protection use in flood prevention; fourth, the relatively low capacity of water resources in supporting high-quality development and the GDP per capita are obstacles to a sustainable development of the basin; fifth, there is still upside potential in the creation and innovation of modern water culture, and the role of water economy and culture in boosting socio-economic development and people's well-being needs to be further enhanced.

4.7 Mississippi River

The Mississippi River, stretching 6,021 km from its source at a tiny stream of Missouri River, the largest tributary of Mississippi River in the Rocky Mountains, is the longest river in North America and the fourth longest river worldwide. With a 58 billion m^3 of average annual discharge and sediment transport capacity of 495 million tons per year, the Mississippi River drains an area of 3.22 million km^2 covering the vast areas of eastern and central U.S. that take up 41 percent of the territory, and flows into the Atlantic Ocean at the Gulf of Mexico.

Located in one of the three major black soil zones of the world and praised as the national treasury of culture and recreation, the Mississippi River and its floodplain are home to over 400 species of wildlife, and more than half of the North American birds use it as their habitats and migratory flyway. A quarter of the fishery resources in the continent is provided by the watershed, and this vital north-south waterway supports a massive industry of water transportation in the U.S.

The RHI of the Mississippi receives an overall score of 80.1 points, thus reaching the grade of "medium high" (Figure 4.7).

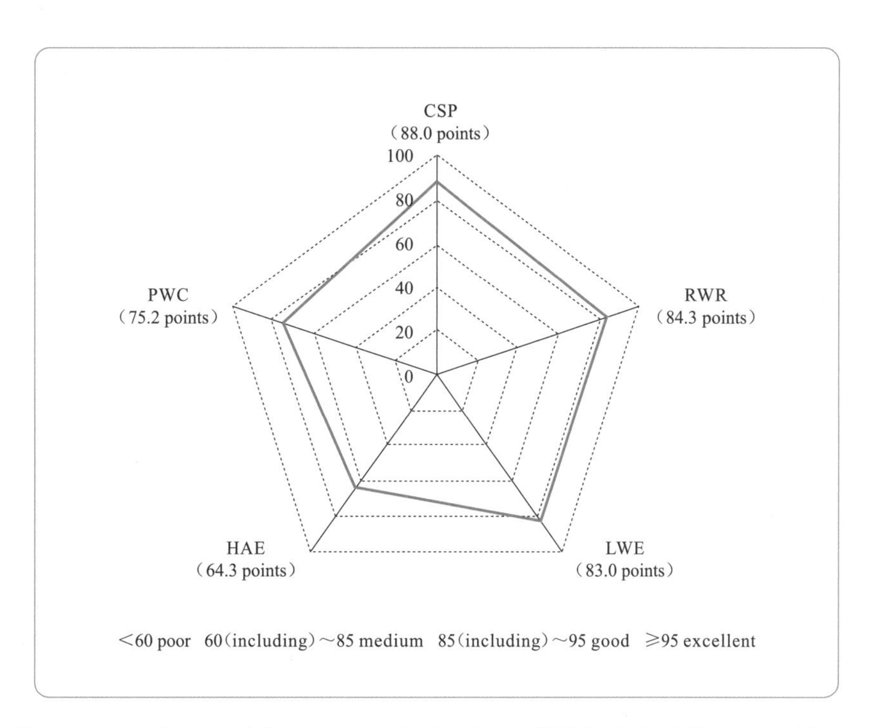

Figure 4.7 Scores of five primary indicators of RHI of the Mississippi River

Capacity of Life and Property Safety Protection (CSP). The CSP of the Mississippi River receives a score of 88.0 points, which is graded as "good". Among the three secondary indicators under the CSP, the Flood-Induced Mortality Rate (FMR) scores 80.0 points and is graded as "medium high"; while the Economic Impact Rate (EIR) and the Flood Disaster Prevention and Adaptation Capacity (PAC) both are graded as "good" with respective scores of 90.0 points and 92.4 points.

Reliability of Water Resources (RWR). The RWR of the Mississippi River receives a score of 84.3 points and is at the grade of "medium high". The Available Water Volume Per Capita (AWP), with an amount of 13,622 m^3 within the watershed, scores 100 points and obtains the grade of "excellent"; the Water Supply Reliability (WSR) and the Capacity for Supporting High-Quality Development (CSD) score 80.3 points and 81.2 points respectively and are both graded as "medium high"; and the Life Satisfaction Index (LSI) scores 79.9 points and is graded as "medium".

Livability of Water Environment (LWE). The LWE of the Mississippi River gets a score of 83.0 points and is graded as "medium". For the four secondary indicators, the Proportion of Water Bodies with Good Water Quality (PGW) scores 72.6 points and is graded as "medium"; the Percentage of Population with Safely Managed Drinking Water Services (PSD) scores 99.0 points, thus attaining an "excellent"; the Urban Wastewater Treatment Rate (WTR) scores 93.3 points and is graded "good"; and the Waterfront Index (WFI) gets a score of 64.4 points, which falls into the grade of "medium low".

Health of Aquatic Ecosystem (HAE). The HAE of the Mississippi River is graded as "medium low" with a score of 64.3 points. Regarding the four secondary indicators, the Variation Index of Eco-Hydrological Process (VIH) is rated as "extremely poor" due to its score of 27.7 points; the River Longitudinal Connectivity Index (LCI) scores 67.4 points and is graded as "medium low"; while the Fish Endangered Index (FEI) reaches the grade of "good" with a score of 93.3 points; and the Sediment Transport Modulus (STM) reaches "medium high" with 82.0 points.

Prosperity of Water-related Culture (PWC). The PWC of the Mississippi River scores 75.2 points and is graded as "medium". Among the three secondary indicators, the Water Culture Protection and Inheritance Index (CPI) scores 69.9 points and is graded as "medium low"; and the Modern Water Culture Creation and Innovation Index (MCI) is graded as "medium" with 76.2 points; finally the Public Awareness and Involvement in Water Governance (PAG) scores 81.3 points and reaches the grade of "medium high".

The results evaluated according to the RHI of the Mississippi demonstrate that: first, overall speaking, the watershed has been well managed in a long-term and systematic manner, particularly on water security, water resources and water environment that are all above the grade of "medium", and has achieved a good public awareness and engagement in water governance; second, the unsatisfactory scores of the VIH and the LCI indicate the noteworthy impact of human activities on natural habitats in this highly-developed watershed; third, there is still a shortage in the inheritance and protection of water heritages.

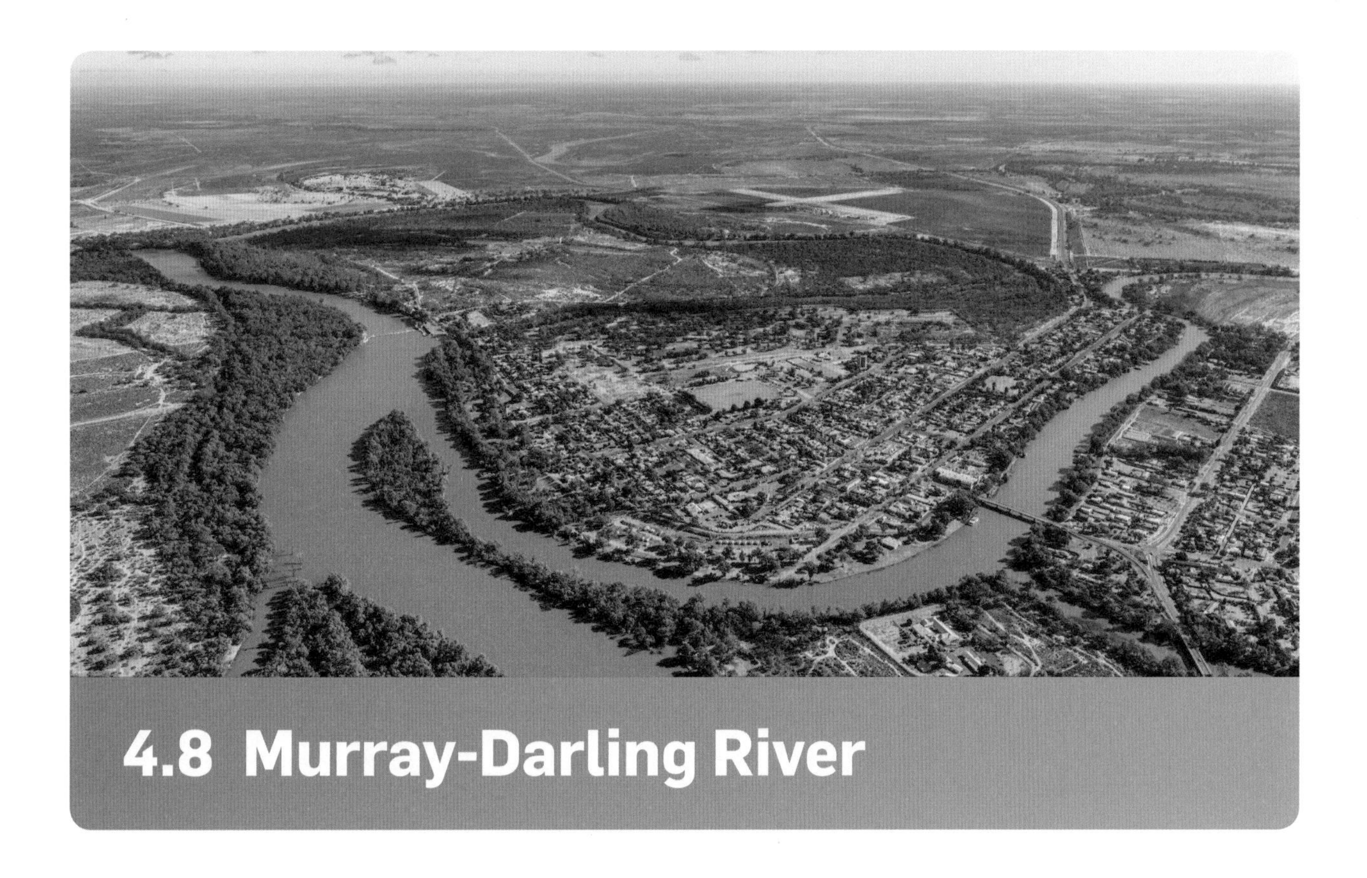

4.8 Murray-Darling River

The Murray-Darling River rises in the southeast of New South Wales, Australia, with a total length of 3,672 km, a drainage area of 1,557,000 km^2 and an average annual discharge of 715 m^3/s and an average annual runoff of 23.6 billion m^3 at its mouth. Flowing roughly west and northwest, the Murray-Darling River flows through the central lowlands southeast of the continent, including southern Queensland, northern Victoria and most of New South Wales, and eventually into the southern Indian Ocean at Encounter Bay.

Covering 14 percent of Australia's land area, the Murray-Darling River is the heartland of the agriculture businesses, and of immense economic and social significance for the country. Over 30 thousand wetlands and floodplains sustain environments for hundreds of animal associations, including more than 50 fish species, 350 bird species, and in particular 95 endangered species. The integrated management of the basin goes back a long way and now has gained an international recognition.

The RHI of the Murray-Darling River receives an overall score of 75.6 points, and is graded as "medium" (Figure 4.8).

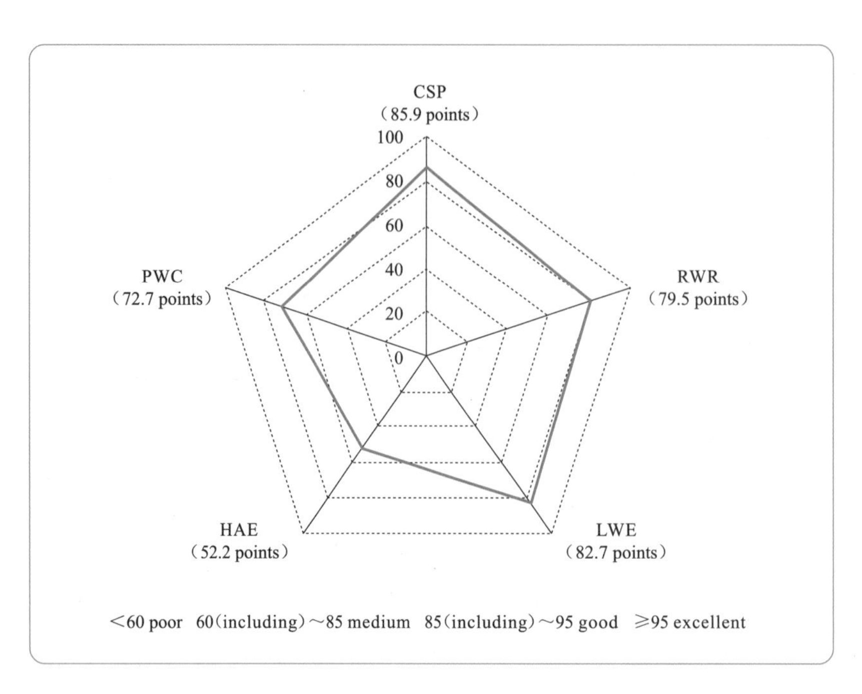

Figure 4.8 Scores of five primary indicators of RHI of the Murray-Darling

Capacity of Life and Property Safety Protection (CSP). The CSP of the Murray-Darling receives a score of 85.9 points, which is graded as "good". Among the three secondary indicators under the CSP, the Flood Disaster Prevention and Adaptation Capacity (PAC) scores 94.7 points and reaches the grade of "good"; the Flood-Induced Mortality Rate (FMR) and the Economic Impact Rate (EIR) both score 80.0 points and are graded as "medium high".

Reliability of Water Resources (RWR). The RWR of the basin receives a score of 79.5 points, which is graded as "medium". For the four secondary indicators, the Available Water Volume Per Capita (AWP), with an amount of 1,239.1 m^3, scores 100.0 points and is graded as "excellent"; the Life Satisfaction Index (LSI) obtains a score of 87.0 points and is graded as "good"; the Water Supply Reliability (WSR) scores 76.8 points and is graded as "medium"; while the Capacity for Supporting High-Quality Development (CSD) scores 57.1 points and thus is graded as "quite poor".

Livability of Water Environment (LWE). The LWE within the basin is at the grade of "medium high" with a score of 82.7 points. Among the four secondary indicators, the Proportion of Water Bodies with Good Water Quality (PGW) scores 75.0 points and is graded as "medium"; the Percentage of Population with Safely Managed Drinking Water Services (PSD) reaches "excellent" by scoring 98.8 points; the Urban Wastewater Treatment Rate (WTR) scores 92.9 points and is graded as "good"; while the Waterfront Index (WFI) falls to the grade of "medium low" with a score of 60 points.

Health of Aquatic Ecosystem (HAE). The HAE of the basin receives 52.2 points and is graded as "quite poor". And for the four secondary indicators under the HAE, the Variation Index of Eco-Hydrological Process (VIH) scores 8.1 points and is graded as "extremely poor"; the River Longitudinal Connectivity Index (LCI) gets a score of 77.2 points, which is graded as "medium"; the Fish Endangered Index (FEI) scores 44.0 points and is graded as "quite poor"; and the Sediment Transport Modulus (STM) scores 86.7 points and reaches "good".

Prosperity of Water-related Culture (PWC). The PWC of the Murray-Darling River gains a points of 72.7 points, which is graded as "medium". For the three secondary indicators, the Water Culture Protection and Inheritance Index (CPI) scores 76.2 points and is graded as "medium"; the Modern Water Culture Creation and Innovation Index (MCI) is at "medium low" by scoring 60.0 points; the Public Awareness and Involvement in Water Governance (PAG) score 80.7 points and is graded as "medium high".

The evaluation results of the RHI of the Murray-Darling River Basin show that: first, the water resources has been greatly developed and fully used, yet its capacity for supporting high-quality development is limited, restricting the regional sustainability; second, the waterfront areas in the cities are under-proportionate, indicating that more efforts should be put in building a livable water environment; third, the variation of eco-hydrological process and the status of endangered fish species are notable, severely disrupting the natural habitats and biological communities within the basin, which is a problem to be solved for protecting river ecosystems in the future.

4.9 Nile River

The Nile River, the longest river in the world, rises in the East African Highlands, has a length of 6,670 km and drains an area of 3.35 million km^2, with an average annual runoff of 2,500 m^3/s. The Nile River basin is bounded by the Arabian Desert to the east, the Libyan Desert to the west and the Nubian Desert to the south, and stretches through countries such as Egypt, Ethiopia, Burundi, Eritrea, the Democratic Republic of the Congo, Kenya, Rwanda, Sudan, South Sudan, Tanzania and Uganda, before draining into the Mediterranean.

The upper Nile River is a tropical rainy region with tremendous flow; despite much runoff lost to evaporation and seepage along the deserts, the Nile River is still the international river that has the longest course and runs through the most African countries in the world. The Nile River is the cradle of ancient Egyptian civilization, and is hailed as the "mother river" and lifeline of Egypt. The ancient Egyptians depended on the Nile for survival and development, and created the brilliant Ancient Egyptian civilization.

The Nile River's RHI gets an overall score of 62.1 points, and is graded as "medium low" (Figure 4.9).

Capacity of Life and Property Safety Protection (CSP). The CSP of the Nile River gets a score of 69.7 points and is graded as "medium low". For the three secondary indicators under this primary one, the Flood-Induced Mortality Rate (FMR) and the Economic Impact Rate (EIR) are both graded as "medium"

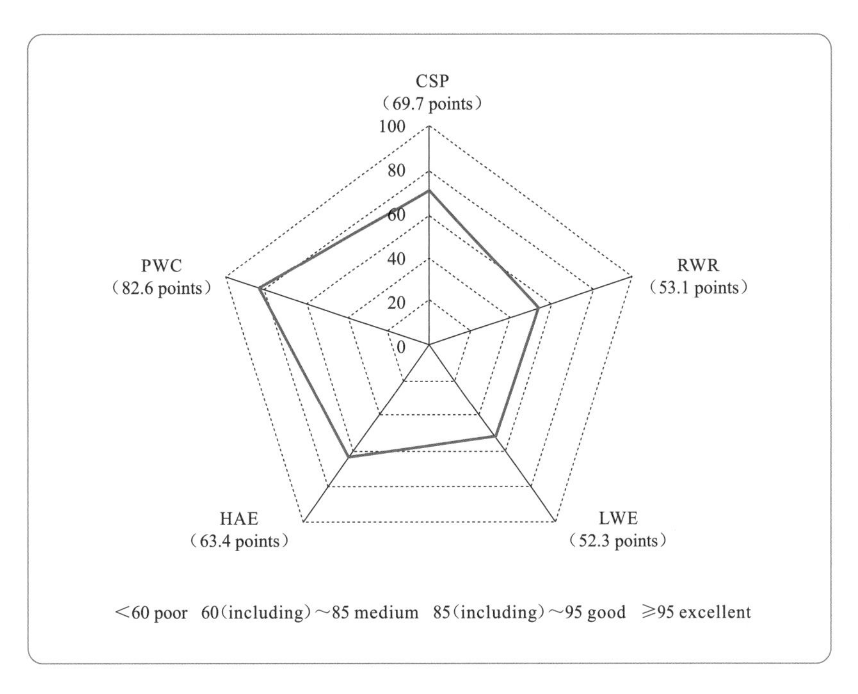

Figure 4.9 Scores of five primary indicators of RHI of the Nile River

with respective scores of 75.0 points and 70.0 points; while the Flood Disaster Prevention and Adaptation Capacity (PAC) scores 65.6 points and is graded as "medium low".

Reliability of Water Resources (RWR). The Nile River's RWR receives a score of 53.1 points and is graded as "quite poor". Regarding the four secondary indicators, namely, the Available Water Volume Per Capita (AWP), the Water Supply Reliability (WSR), the Capacity for Supporting High-Quality Development (CSD) and the Life Satisfaction Index (LSI) are all graded as "quite poor" with scores of 52.0 points, 56.0 points, 51.2 points and 52.4 points separately.

Livability of Water Environment (LWE). The LWE of the Nile River obtains a score of 52.3 points, which is graded as "quite poor". Among the four secondary indicators, the Proportion of Water Bodies with Good Water Quality (PGW) is at the grade of "medium low" by scoring 63.6 points; the Percentage of Population with Safely Managed Drinking Water Services (PSD) and the Urban Wastewater Treatment Rate (WTR) score 34.4 points and 32.3 points respectively, which both fall into the grade of "quite poor"; while the Waterfront Index (WFI) scores 82.1 points and reaches "medium high".

Health of Aquatic Ecosystem (HAE). The HAE of the Nile River scores 63.4 points and is graded as "medium low". For the four secondary indicators under the HAE, the Variation Index of Eco-Hydrological Process (VIH) scores 17.8 points and is graded as "extremely poor"; the River Longitudinal Connectivity Index (LCI) is at the grade of "medium high" with a score of 80.5 points; the Fish Endangered Index (FEI) scores 93.3 points and is graded as "good"; and the Sediment Transport Modulus (STM) is graded as "medium" by scoring 77.3 points.

Prosperity of Water-related Culture (PWC). The PWC of the Nile River gains a score of 82.6 points and is graded as "medium high". For the three secondary indicators, the Water Culture Protection and Inheritance Index (CPI) and the Modern Water Culture Creation and Innovation Index (MCI) are both graded as "good" with respective scores of 85.6 points and 92.6 points; while the Public Awareness and Involvement in Water Governance (PAG) scores 68.5 points and is graded as "medium low".

The evaluation results of the Nile River's RHI demonstrate that: first, flood defense in the basin is a key issue for enhancing life and property protection; second, the basin has not yet guaranteed a reliable water supply, the capacity of water resources in supporting high-quality development remains to be advanced; third, the water environment of the riverine areas is poor, neither are the proportion of good-quality water resources and the urban sewage treatment; fourth, there is lack of impetus for water culture innovation in modern times, and more work should be done in public awareness raising and participation in water management.

4.10 Rhine River

The Rhine River, the longest river in western Europe, stretching 1,320 kilometers from its source of the northern Alps in Graubünden, Switzerland to its mouth on the North Sea in Rotterdam, the Netherlands. Flowing through Switzerland, Italy, Austria, Germany, France, Liechtenstein, Belgium, Luxembourg and the Netherlands. It has a drainage area of 161,000 km^2 and an average annual runoff of 82.1 billion m^3 in Aswan.

Owing to its favorable natural conditions with temperate maritime climate, the Rhine River is of enormous economic and social importance for Europe. Recognized as a "golden waterway", this vast basin has an extensive and well-connected traffic network, thus being the busiest river across the globe. Along the river, the riparian plain is highly urbanized where agriculture and industries thrive.

The RHI of the Rhine River gets an overall score of 86.6 points, and is graded as "happy"(Figure 4.10).

Capacity of Life and Property Safety Protection (CSP). The CSP of the Rhine River receives a score of 95.1 points, which is graded as "excellent". For the three secondary indicators under the CSP, the Flood-Induced Mortality Rate (FMR) reaches the grade of "good" by scoring 90.0 points; the Economic Impact Rate (EIR) gains a full score of 100.0 points that is graded as "excellent"; the same grade goes to the Flood Disaster Prevention and Adaptation Capacity (PAC) that scores 95.3 points.

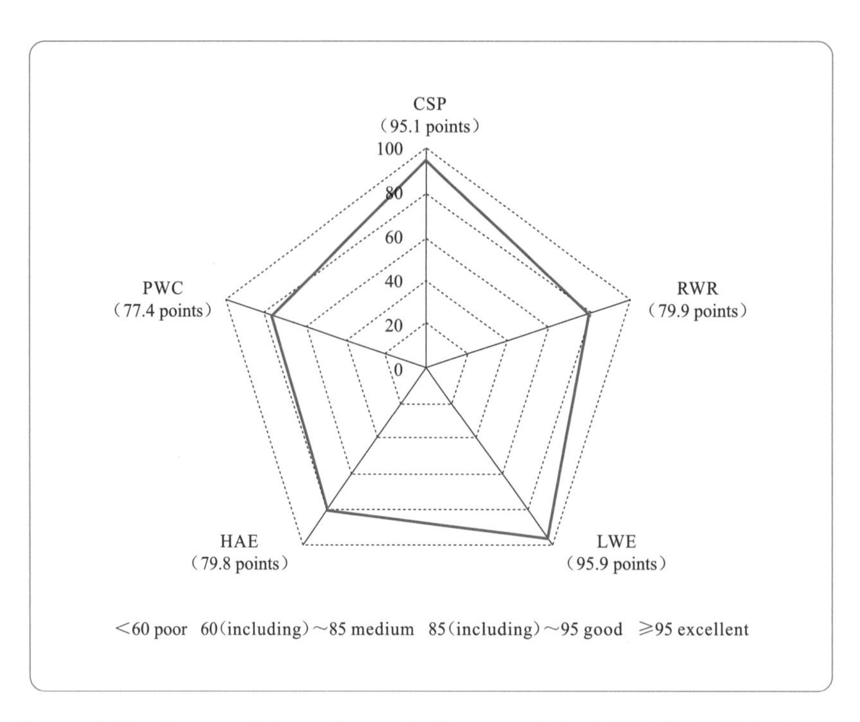

Figure 4.10 Scores of five primary indicators of the RHI of the Rhine River

Reliability of Water Resources (RWR). The RWR of the Rhine River is at the grade of "medium" by scoring 79.9 points. For the four secondary indicators, the Available Water Volume Per Capita (AWP) gets a score of 81.5 points and is graded as "medium high"; the Water Supply Reliability (WSR) scores 54.5 points, which falls to the grade of "quite poor"; the Capacity for Supporting High-Quality Development (CSD) scores 100.0 points and is graded as "excellent"; and the Life Satisfaction Index (LSI) scores 89.1 points and is graded as "good".

Livability of Water Environment (LWE). The LWE of the Rhine River has a score of 95.9 points and reaches "excellent". Among the four secondary indicators, the Proportion of Water Bodies with Good Water Quality (PGW), the Percentage of Population with Safely Managed Drinking Water Services (PSD) and the Urban Wastewater Treatment Rate (WTR) all reach "excellent" by scoring 96.4 points, 99.0 points and 96.0 points respectively; and the Waterfront Index (WFI) scores 90.4 points and is graded as "good".

Health of Aquatic Ecosystem (HAE). The Rhine River's HAE scores 79.8 points and is graded as "medium". For the four secondary indicators, the Variation Index of Eco-Hydrological Process (VIH) scores 51.0 points and thus is graded as "quite poor"; the River Longitudinal Connectivity Index (LCI) reaches "excellent" with a score of 96.6 points; the Fish Endangered Index (FEI) and the Sediment Transport Modulus (STM) both reach "good" with respective scores of 94.7 points and 85.7 points.

Prosperity of Water-related Culture (PWC). The PWC of the Rhine River receives a score of 77.4

points and is at the grade of "medium". Among the three secondary indicators, the Water Culture Protection and Inheritance Index (CPI) scores 81.5 points and is graded as "medium high"; the Modern Water Culture Creation and Innovation Index (MCI) is graded as "medium low" by scoring 60.8 points; while the Public Awareness and Involvement in Water Governance (PAG) scores 88.6 points and is graded as "good".

The evaluation results of the Rhine River's RHI reflect that: first, the basin has not yet offered a reliable water supply or a strong support in high-quality development; second, the integrity of the Rhine ecosystem is constrained by the variation of eco-hydrological process to a large extent; third, the creation and innovation in modern water culture nearby the basin is remarkably insufficient, and the economic and cultural significance of water shall be enhanced for socio-economic development and people's well-being improvement.

4.11 Saint Lawrence River

The Saint Lawrence River, a hydrographic system of east-central North America, extends 3,058 km and drains an area of 1.054 million km^2; the river features an abundant and stable volume of water, low sediment content and an average annual flow volume of 10,540 m^3/s at its mouth. The St. Lawrence originates in Lake Ontario, flows through Montreal, Canada and then leads into the Atlantic Ocean at the Gulf of St. Lawrence in Gaspé. It is the waterway linking the Great Lakes to the sea.

Possessing fertile and extensive plains with diverse landscape, the basin is rich in various natural resources including water, minerals, forests and animal species. The riverine region, with pleasant scenery, is a world-renowned tourist resort where the Isle Royale National Park and the Niagara Falls are located. The shipping industry thrives there as the river can carry ocean vessels directly to the Great Lakes, making it one of the rivers with highest commercial value in the Americas. The convenience in river shipping also fosters North America's earliest industrial center that is scattered with towns and large-scale agricultural and industrial plants, making great contributions to the economy of Canada and the U.S.

The RHI of the St. Lawrence River gets an overall score of 84.6 points, and is graded as "medium high" but quite close to "happy"(Figure 4.11).

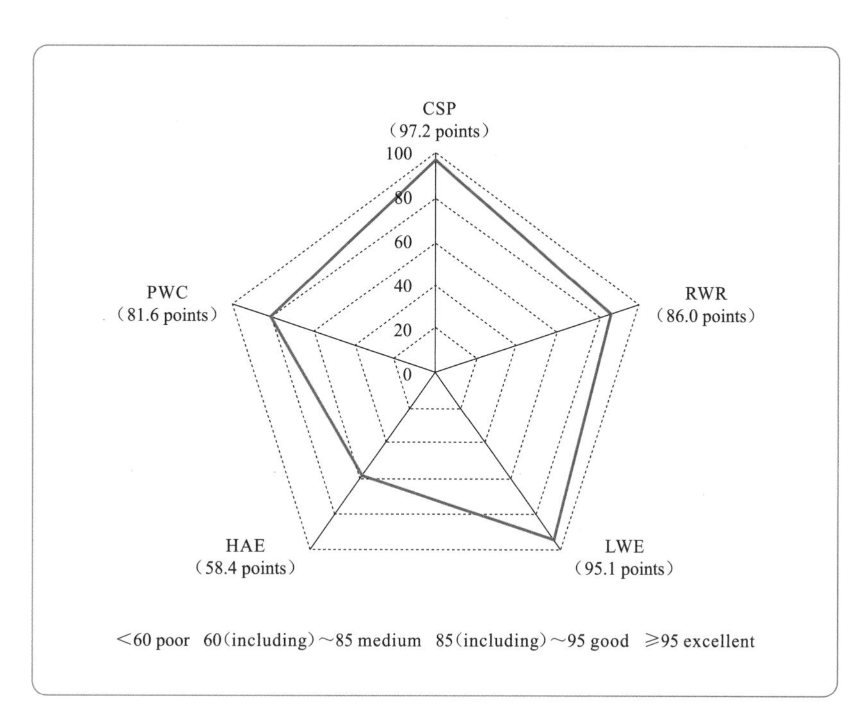

Figure 4.11 Scores of five primary indicators of RHI of the St. Lawrence River

Capacity of Life and Property Safety Protection (CSP). The CSP of the St. Lawrence River receives a score of 97.2 points and reaches "excellent". Among the three secondary indicators under the CSP, due to low death toll and minimal economic loss caused by flood, its Flood-Induced Mortality Rate (FMR) and the Economic Impact Rate (EIR) both score 100.0 points and are graded as "excellent"; while the Flood Disaster Prevention and Adaptation Capacity (PAC) gets a score of 92.9 points that is graded as "good".

Reliability of Water Resources (RWR). The RWR of the river receives a score of 86.0 points and is at the grade of "good". For the four secondary indicators, the Available Water Volume Per Capita (AWP), with an amount of 20,466 m^3, gets a full score of 100.0 points and reaches "excellent"; the Water Supply Reliability (WSR) and the Capacity for Supporting High-Quality Development (CSD) are both graded as "medium high" with respective scores of 80.3 points and 80.5 points; and the Life Satisfaction Index (LSI) scores 87.2 points and is graded as "good".

Livability of Water Environment (LWE). The LWE of the St. Lawrence River gains a score of 95.1 points and reaches "excellent". Among the four secondary indicators, the Proportion of Water Bodies with Good Water Quality (PGW) and the Percentage of Population with Safely Managed Drinking Water Services (PSD) both score over 95.0 points, which are graded as "excellent"; the Urban Wastewater Treatment Rate (WTR) and the Waterfront Index (WFI) both reach "good" by scoring 90.2 points and 94.1 points respectively.

Health of Aquatic Ecosystem (HAE). The HAE of the river scores 58.4 points, which falls to the grade of “quite poor”. Among the four secondary indicators, the Variation Index of Eco-Hydrological Process (VIH) and the River Longitudinal Connectivity Index (LCI) scores 34.4 points and 58.7 points respectively, thus being graded as “quite poor”; the Sediment Transport Modulus (STM) scores 61.1 points and is graded as “medium low”; while the Fish Endangered Index (FEI) reaches “good” with a score of 90.7 points.

Prosperity of Water-related Culture (PWC). The PWC of the basin scores 81.6 points and is graded as “medium high”. For the three secondary indicators, the Water Culture Protection and Inheritance Index (CPI) gets a score of 66.0 points and is graded as “medium low”; the Modern Water Culture Creation and Innovation Index (MCI) scores 96.8 points and reaches “excellent”; and the Public Awareness and Involvement in Water Governance (PAG) scores 87.3 points and is graded as “good”.

The RHI-based results of the St. Lawrence indicate several problems: first, despite the abundance of water resources, the water supply is not secure enough and its capacity for supporting the high-quality development of the countries and the regions along the river is inadequate; second, the basin has not yet attained a healthy hydro-ecology with some imbalances, although the endangered fishes have been well protected and reach “good”, the scores of other three indicators are lower or much lower than the world average; third, limited by historical reasons, the protection and inheritance of historical water culture are insufficient, the innovative culture and social governance, however, level up the overall grade to “medium”.

4.12 Thames River

The Thames River, or River Thames, is 346 km long and a chief river in the United Kingdom. Stemming from the Cotswold Hills of southwestern England, its basin covers an area of 13,000 km^2. The river starts from west and head eastward to Oxford, where it moves southeasterly and turns to northeast after passing through Reading, then changes its course to east again in Windsor toward London, eventually entering the North Sea near Southend. Its annual average flow measures 65.8 m^3/s and average annual runoff measures 1.89 billion m^3.

The Thames River has the world's most loaded waterway traffic in urban area, and its basin is served as an economic powerhouse that benefits river navigation. Identified as a landmark of London, along the riverbank there is dotted with several main buildings of the city and many popular sites such as Eton College, Oxford Street, Henley-on-Thames, royal Windsor Castle, etc. The Thames had been a "dead river" because of pollution since the Industrial Revolution. After years of treatment, the river, which was once declared "biologically dead", is now clear and reinvigorated with a good ecological environment, providing habitat for a variety of aquatic creatures such as seals. The improvement of the Thames is one of the representative cases of river pollution control in the context of urbanization and industrialization.

The RHI of the Thames scores 81.9 points, which is graded as "medium high" (Figure 4.12).

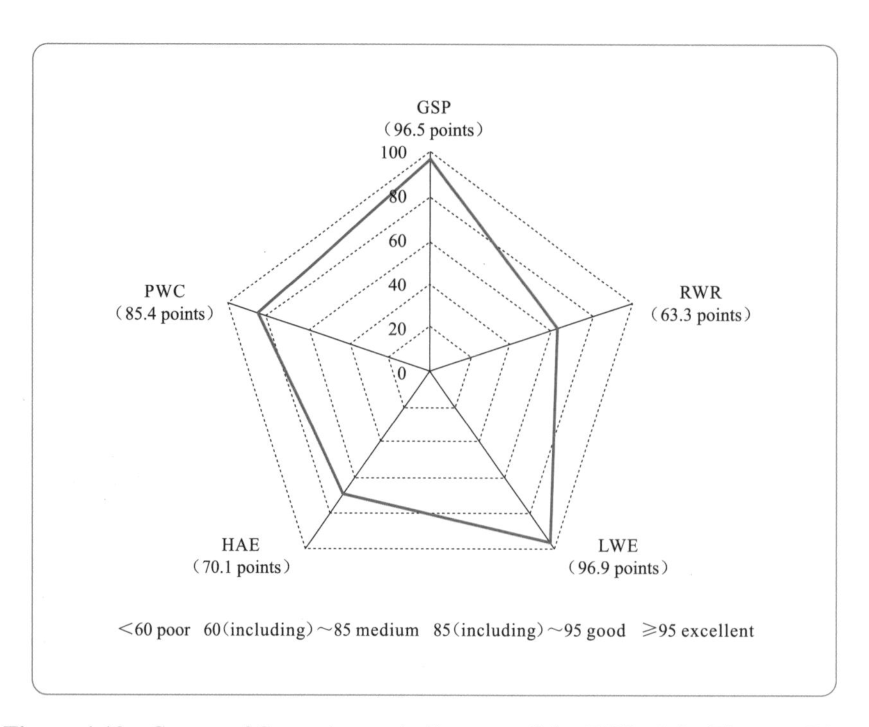

Figure 4.12 Scores of five primary indicators of the RHI of the Thames River

Capacity of Life and Property Safety Protection (CSP). The CSP of the Thames River receives a score of 96.5 points and is graded as "excellent". Among the three secondary indicators under this primary one, the Flood-Induced Mortality Rate (FMR) gains a full score of 100.0 points, thus reaching "excellent"; the Economic Impact Rate (EIR) and the Flood Disaster Prevention and Adaptation Capacity (PAC) reach "excellent" as well and both score 95.0 points.

Reliability of Water Resources (RWR). The RWR of the Thames River scores 63.3 points, which is graded as "medium low". For the four secondary indicators, the Available Water Volume Per Capita (AWP) gets a score of 24.6 points, which falls to the grade of "extremely poor"; the Water Supply Reliability (WSR) scores 56.0 points and is graded as "quite poor"; the Capacity for Supporting High-Quality Development (CSD) scores 80.1 points and is graded as "medium high"; and the Life Satisfaction Index (LSI) scores 86.4 points and reaches "good".

Livability of Water Environment (LWE). The LWE of the Thames River reaches "excellent" by scoring 96.9 points. Among the four secondary indicators, the Proportion of Water Bodies with Good Water Quality (PGW) receives a score of 90.0 points that is graded as "good"; the Percentage of Population with Safely Managed Drinking Water Services (PSD), the Urban Wastewater Treatment Rate (WTR) and the Waterfront Index (WFI) all reach "excellent" with respective scores of 99.9 points, 99.5 points and 100.0 points.

Health of Aquatic Ecosystem (HAE). The HAE of the river scores 70.1 points and is graded as "medium". Regarding the four secondary indicators, the Variation Index of Eco-Hydrological Process (VIH) scores 23.2 points and is graded as "extremely poor"; while the River Longitudinal Connectivity Index (LCI) gains a score of 99.7 points and reaches "excellent"; the Fish Endangered Index (FEI) scores 92.0 points, which is graded as "good"; and the Sediment Transport Modulus (STM) has a score of 79.4 points and is graded as "medium".

Prosperity of Water-related Culture (PWC). The PWC of the basin scores 85.4 points and is graded as "good". Among the three secondary indicators, the Water Culture Protection and Inheritance Index (CPI) is at the grade of "medium" by scoring 78.0 points; the Modern Water Culture Creation and Innovation Index (MCI) and Public Awareness and Involvement in Water Governance (PAG) are both graded as "good" with scores of 89.2 points and 91.4 points respectively.

The results evaluated from the RHI underline two issues of the river: first, the availability of water resources per person and the reliability of water use are both inadequate to support the basin's high-quality development; second the notable change in eco-hydrological process becomes a key factor affecting the hydro-ecological health.

4.13 Volga River

Stretching 3,645 kilometers long, **the Volga River** is the longest river in Europe and ranks the world's longest inland river. It rises in the marshland of the Valdai Hills west of the East European Plain, also known as the Russian Plain, and continues passing through three geographic zones until draining into the Caspian Sea. The river drains an area of 1.38 million km^2, and discharges an average of 255 billion m^3 of water per year. The Volga River is joined to and connected the Baltic Sea, the White Sea, the Sea of Azov and the Black sea, thus being navigable with virtually the entire waterway system of Eastern Europe.

Crowned the historic cradle of Russia, the Volga River basin is populated by dense forest with thriving agriculture. Its integrated waterway network facilitates the river navigation in the region. Also, the Volga Economic Region is a major industrial and agricultural center in the Russian Federation. However, the development of the Volga has also produced certain ecological impacts, with reservoir construction affecting fish reproduction in the river basin and the discharge of wastewater from industry and agriculture causing pollution to the water environment.

The RHI of the Volga River receives an overall score of 79.0 points, which is graded as "medium" (Figure 4.13).

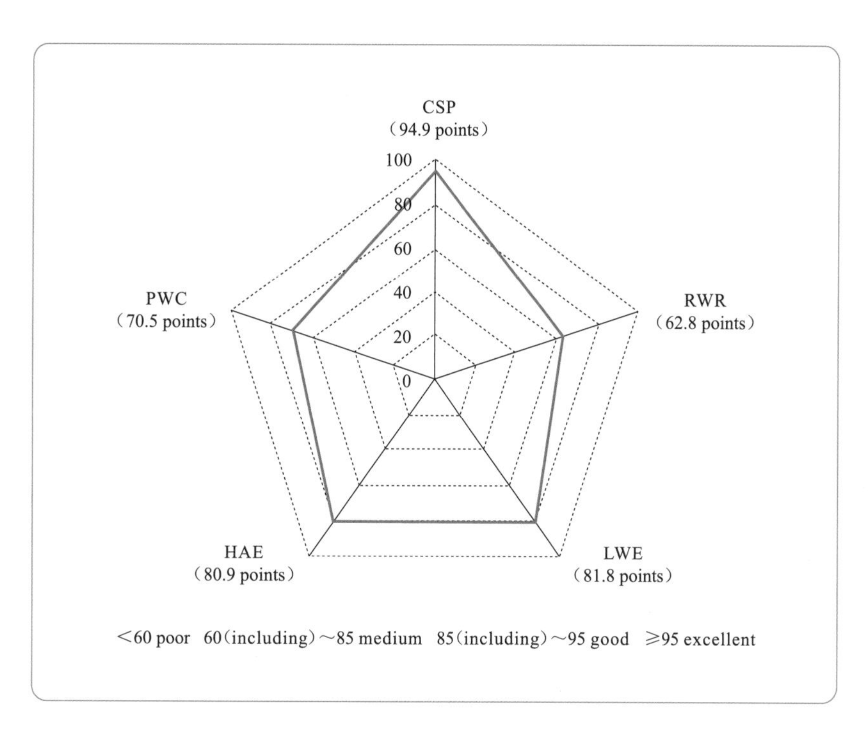

Figure 4.13 Scores five primary indicators of RHI of the Volga River

Capacity of Life and Property Safety Protection (CSP). The CSP of the Volga River receives a score of 94.9 points and is graded as "good" approaching to "excellent". Among the three secondary indicators under the CSP, the Flood-Induced Mortality Rate (FMR) and the Economic Impact Rate (EIR) both gain a full score of 100.0 points, thus being graded as "excellent"; and the Flood Disaster Prevention and Adaptation Capacity (PAC) scores 87.2 points and is graded as "good".

Reliability of Water Resources (RWR). The RWR of the Volga River is at the grade of "medium low" by scoring 62.8 points. For the four secondary indicators, the Available Water Volume Per Capita (AWP) receives a score of 92.5 points and is graded as "good"; the Life Satisfaction Index (LSI) and the Capacity for Supporting High-Quality Development (CSD) scores 65.6 points and 64.2 points respectively and are both graded as "medium low"; the Water Supply Reliability (WSR) scores 39.5 points, which falls to the grade of "quite poor".

Livability of Water Environment (LWE). The Volga River's LWE gets a score of 81.8 points and is graded as "medium high". Among the four secondary indicators, the Proportion of Water Bodies with Good Water Quality (PGW) scores 95.0 points and reaches "excellent"; the Urban Wastewater Treatment Rate (WTR) scores 89.0 points, which is graded as "good"; the Percentage of Population with Safely Managed Drinking Water Services (PSD) scores 76.1 points and is graded as "medium"; and the Waterfront Index (WFI) is at "medium low" with a score of 63.2 points.

Health of Aquatic Ecosystem (HAE). The HAE of the Volga River scores 80.9 points and is graded as “medium high”. Regarding the four secondary indicators under this primary one, the Sediment Transport Modulus (STM) scores 98.6 points and reaches “excellent”; the Fish Endangered Index (FEI) scores 92.0 points that is graded as “good”; the River Longitudinal Connectivity Index (LCI) is at the grade of “medium high” by scoring 84.7 points; the Variation Index of Eco-Hydrological Process (VIH) scores 55.6 points and is graded as “quite poor”.

Prosperity of Water-related Culture (PWC). The PWC of the Volga River receives a score of 70.5 points, which is graded as “medium”. For the three secondary indicator, the Water Culture Protection and Inheritance Index (CPI) and the Public Awareness and Involvement in Water Governance (PAG) both score 75.0 points and are graded as “medium”; while the Modern Water Culture Creation and Innovation Index (MCI) scores 60.0 points and is graded as “medium low”.

The evaluation results of the Volga River’s RHI reflect the following issues: first, water resources in the basin are still unable to support the high-quality development, and the secure water supply is a matter of concern as well; second, the variation of eco-hydrological process is one of the main indicators that brings down the river’s RHI grade; third, there is so much potential for the creation and innovation in modern water culture, and a gap still exists between current status of water culture and people’s demand; fourth, the riverfront areas shall be incorporated into urban planning to build a more livable water environment.

4.14 Yangtze River

The largest river in China, **the Yangtze River** lies in the south-central China belonging to a subtropical monsoon climate. From its source on the Plateau of Tibet to its mouth on the East China Sea, the Yangtze River main stream traverses 11 provincial-level administrative regions including Qinghai, Tibet, Sichuan, Yunnan, Chongqing, Hubei, Hunan, Jiangxi, Anhui, Jiangsu, and Shanghai, for a distance of 6,397 km. With an average annual average runoff of 976 billion m^3, the Yangtzes' tributaries extend to 8 provinces or autonomous regions including Gansu, Shaanxi, Guizhou, Henan, Guangxi, Guangdong, Fujian and Zhejiang. The Yangtze River drains an estimated area of 1.784 million km^2, which occupies 18.8 percent of the land area of China. There are 437 Yangtze tributaries measuring more than 1,000 km^2 in drainage area, 49 of which are over 10 thousand km^2, and 8 over 80 thousand km^2. The Yangtze Region consists of a large number of lakes, many of which are in the middle and lower course.

Drawing upon its privileged natural environment, the Yangtze River Basin is among the most economically developed areas and of considerable social and economic significance for the country. The basin is characterized by temperate climate, plentiful precipitation, fertile land and enormous solar resources, thus historically being China's important center for agriculture and crop production. Six plains situated on the basin including Chengdu Plain, Jianghan Plain, Dongting Lake, Poyang Lake, Chaohu Lake, and Taihu Lake, are main production sites for crop, cotton and oil. Within the basin, there are 15 megacities and 89 cities above prefecture level that take up 31.8 percent of the nationwide total. Five urban economic spheres have been developed relying on the river as well, consisting of the Yangtze Delta Region, the Wanjiang Urban Belt, the Wuhan City Circle, the Chang-Zhu-Tan City Group and the Chengdu-Chongqing Economic Zone.

The Yangtze River gets an overall score of 80.8 points in RHI, which is graded as "medium high" (Figure 4.14).

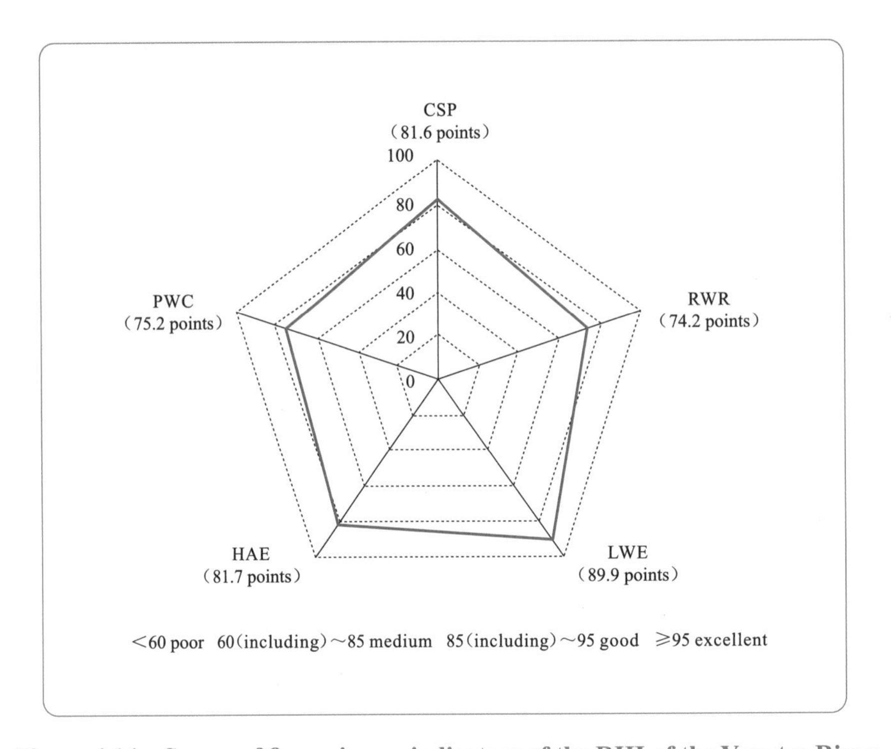

Figure 4.14 Scores of five primary indicators of the RHI of the Yangtze River

Capacity of Life and Property Safety Protection (CSP). The CSP of the Yangtze River receives a score of 81.6 points, which is graded as "medium high". Among the three secondary indicators under the CSP, the Flood-Induced Mortality Rate (FMR) scores 85.0 points and reaches "good"; the Flood Disaster Prevention and Adaptation Capacity (PAC) scores 84.1 points and is graded as "medium high"; and the Economic Impact Rate (EIR) is rated as "medium" with a score of 75.0 points.

Reliability of Water Resources (RWR). The RWR of the river gets a score of 74.2 points that is graded as "medium". Among the four indicators under the RWR, the Available Water Volume Per Capita (AWP) and the Water Supply Reliability (WSR) scores 81.2 points and 82.9 points respectively and both are graded as "medium high"; while the Life Satisfaction Index (LSI) and the Capacity for Supporting High-Quality Development (CSD) are at the grade of "medium low" with respective scores of 66.9 points and 65.6 points.

Livability of Water Environment (LWE). The LWE of the Yangtze River receives a score of 89.9 points and is graded as "good". The four secondary indicators, namely, the Urban Wastewater Treatment Rate (WTR), the Waterfront Index (WFI), the Percentage of Population with Safely Managed Drinking Water Services (PSD), and the Proportion of Water Bodies with Good Water Quality (PGW) all reach "good" by scoring 94.6 points, 94.1 points, 88.4 points and 85.6 points separately.

Health of Aquatic Ecosystem (HAE). The basin's HAE scores 81.7 points and is graded as "medium high". For the four secondary indicators, the Sediment Transport Modulus (STM) scores 98.6 points and reaches "excellent"; the Fish Endangered Index (FEI) gets a scores of 90.7 points, thus being graded as "good"; the River Longitudinal Connectivity Index (LCI) scores 76.5 points, which is graded as "medium"; the Variation Index of Eco-Hydrological Process (VIH) scores 65.9 points and is graded as "medium low".

Prosperity of Water-related Culture (PWC). The PWC of the Yangtze River basin scores 75.2 points and is graded as "medium". For the three secondary indicators, the Water Culture Protection and Inheritance Index (CPI) is at the grade of "medium high" by getting a score of 82.8 points; the Public Awareness and Involvement in Water Governance (PAG) scores 74.2 points, which is graded as "medium"; and the Modern Water Culture Creation and Innovation Index (MCI) scores 66.0 points and is graded as "medium low".

The RHI-based results of the Yangtze River showcase that: first, the basin is still economically vulnerable to flood disasters, thus undermining its protection to life and property of the riverine regions; second, the water resources are not capable of pillaring the high-quality development and people's welfare, which are the leading factors in providing sustainable water resources; third, the significant variation of eco-hydrological process threatens the hydro-ecological health of the Yangtze basin; fourth, the creation and innovation of modern water culture still have lots of room to improve, in a bid to fulfill people's increasing needs for water culture.

4.15 Yellow River

The Yellow River is the second largest river in China, with a length of 5,464 km, a drainage area of 795,000 km^2, and an average annual runoff of 58 billion m^3. The river rises in the Yueguzonglie Basin at an elevation of 4,500 m at the northern foot of the Bayan Har Mountains on the Qinghai-Tibet Plateau. It flows from west to east through nine provinces (autonomous regions) including Qinghai, Sichuan, Gansu, Ningxia, Inner Mongolia, Shanxi, Shaanxi, Henan and Shandong, before draining into the Bohai Sea in Kenli District, Dongying City, Shandong Province.

With a low amount of precipitation, which primarily occurs in July and August, and dry winters and springs, the Yellow River basin is dominated by dry farming. As the middle course of the Yellow River flows through the Loess Plateau, it entrains a large amount of silt and sand, making the river the sandiest in the world. The Yellow River basin is the main birthplace of Chinese civilization, and is dubbed by the Chinese as the "mother river". For more than 3,000 years, the Yellow River basin has been the political, economic and cultural center of China, which has given birth to various cultures. It is also an important economic zone in China for its numerous industrial bases of energy, chemicals, raw materials, etc.

The RHI of the Yellow River gets an overall score of 78.8 points, and is graded as "medium" (Figure 4.15).

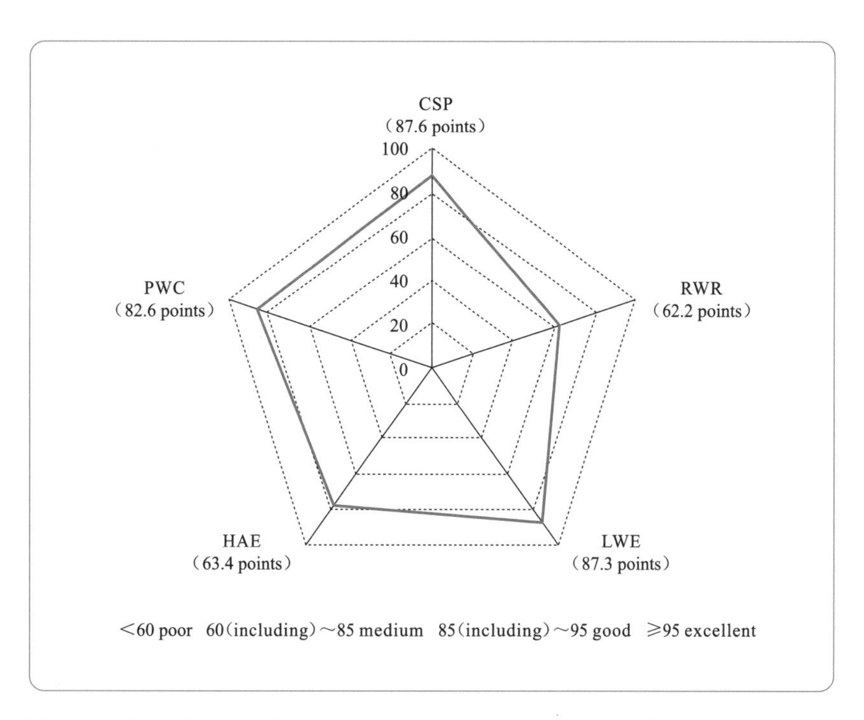

Figure 4.15 Scores of five primary indicators of RHI of the Yellow River

Capacity of Life and Property Safety Protection (CSP). The CSP of the Yellow River receives a score of 87.6 points and is graded as "good". Among the three secondary indicators, the Flood-Induced Mortality Rate (FMR) reaches "excellent" by scoring 95.0 points; the Economic Impact Rate (EIR) scores 85.0 points and is graded as "good"; and the Flood Disaster Prevention and Adaptation Capacity (PAC) is at "medium high" with a score of 84.1 points.

Reliability of Water Resources (RWR). The RWR of the Yellow River scores 62.2 points, which is graded as "medium low". For the four secondary indicators under this primary one, the Available Water Volume Per Capita (AWP) and the Capacity for Supporting High-Quality Development (CSD) get respective scores of 44.8 points and 42.9 points, which both fall to the grade of "quite poor"; the Water Supply Reliability (WSR) scores 89.6 points and is graded as "good"; the Life Satisfaction Index (LSI) scores 62.5 points and is graded as "medium low".

Livability of Water Environment (LWE). The LWE of the river gains a score of 87.3 points that is graded as "good". Regarding the four secondary indicators, the Proportion of Water Bodies with Good Water Quality (PGW) scores 82.2 points and is graded as "medium high"; the Percentage of Population with Safely Managed Drinking Water Services (PSD) reaches "good" by scoring 91.3 points; the Urban Wastewater Treatment Rate (WTR) scores 97.9 points and thus is graded as "excellent"; while the Waterfront Index (WFI) scores 78.1 points, which is graded as "medium".

Health of Aquatic Ecosystem (HAE). The HAE of the river receives a score of 63.4 points and is graded as "medium". For the four secondary indicators, the Variation Index of Eco-Hydrological Process (VIH) scores 56.0 points and is graded as "quite poor"; the River Longitudinal Connectivity Index (LCI) is at "medium" by getting a score of 75.9 points; the Fish Endangered Index (FEI) reaches "good" with a score of 89.3 points; and the Sediment Transport Modulus (STM) approaches to "excellent" with a score of 94.7 points.

Prosperity of Water-related Culture (PWC). The basin's PWC scores 82.6 points, which is close to "good". Among the three secondary indicators, the Water Culture Protection and Inheritance Index (CPI) and the Modern Water Culture Creation and Innovation Index (MCI) both reach "good" by getting scores of 88.9 points and 90.9 points separately; while the Public Awareness and Involvement in Water Governance (PAG) scores 72.6 points and is graded as "medium".

The RHI results of the Yellow River reveal that: firstly, the water resources of the river are not rich in nature and have been heavily exploited with a low availability per capita, therefore being incapable of supporting the high-quality development; secondly, the results underscore a major issue to be addressed in water environment enhancement within the basin, that is, the heavy pollution in its tributaries; thirdly, the variation of eco-hydrological process is alarming in maintaining the healthy water ecology; fourthly, the branding of its water culture and the water landscape need to be further promoted.

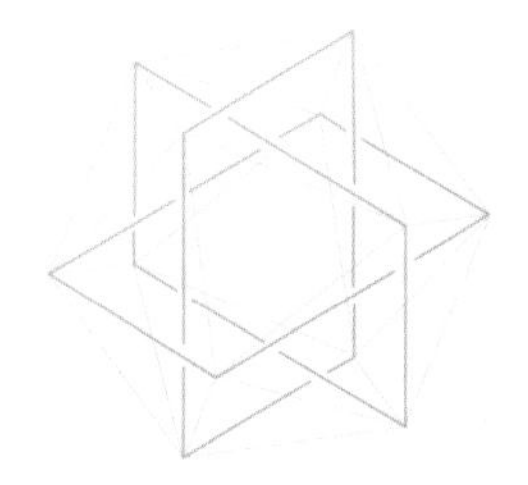

Chapter 5

Comparative Analysis

5.1 Scores and Grades of Happiness of World's Rivers

Our evaluation of the world's 15 rivers using the River Happiness Index vastly yields "medium" of happiness. Among them, the Rhine is graded as "happy" with the highest score of 86.6 points, while the other 14 rivers are all graded as "medium". Specifically, the happiness of the St. Lawrence River, the Thames River, the Colorado River, the Yangtze River and the Mississippi River are graded as "medium high"; the happiness of the Danube River, the Volga River, the Yellow River, the Murray-Darling River, the Amazon River and the Congo River are graded as "medium"; and the happiness of the Euphrates River, the Ganges River and the Nile River are graded as "medium low" (Figure 5.1).

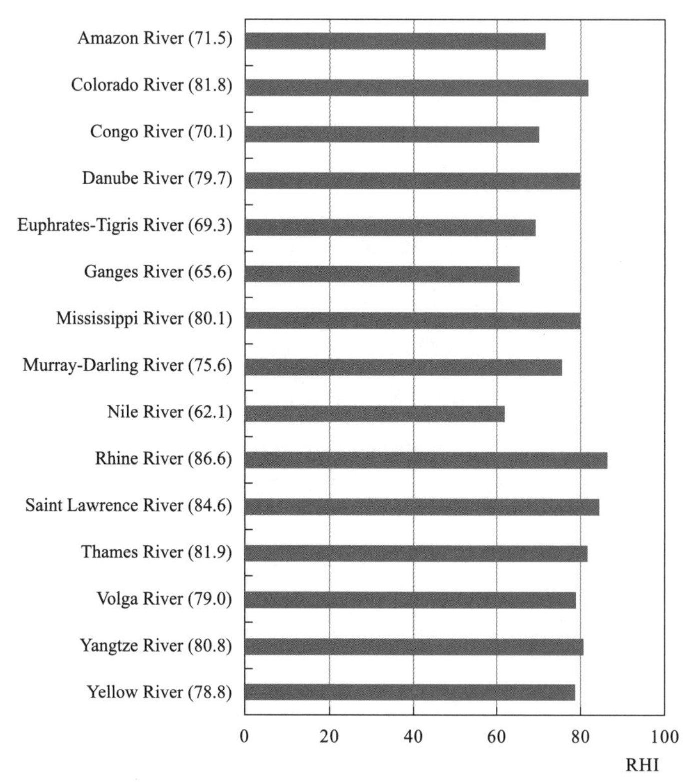

Figure 5.1 A bar diagram of the RHI of evaluated rivers

5.2 Capacity of Life and Property Safety Protection (CSP)

Most of the 15 rivers score “medium” or above in indicators under the Capacity of Life and Property Safety Protection (CSP) (Figure 5.2), meaning the people living along the rivers are well protected and feel generally secure.

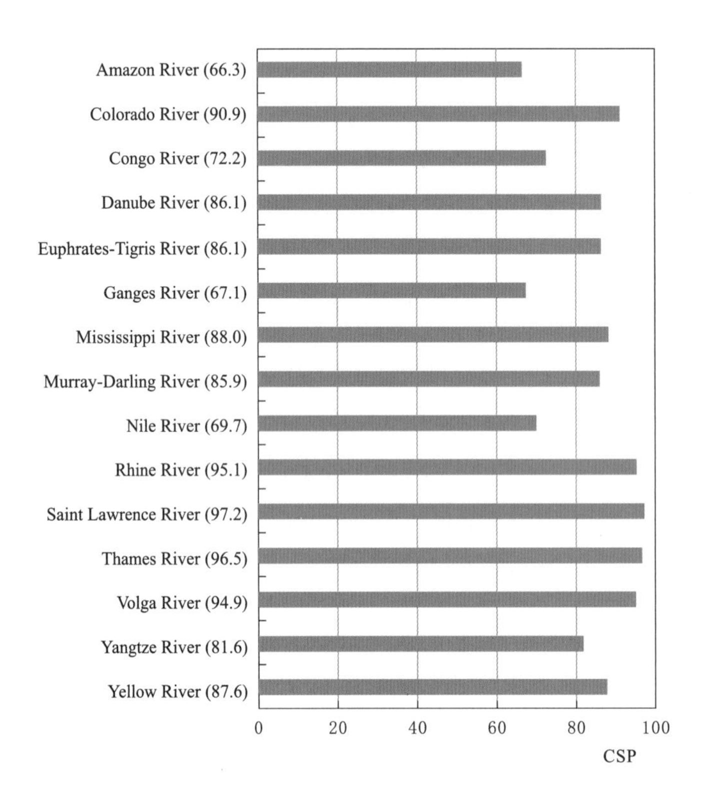

Figure 5.2 Evaluation results of CSP

Note: The scores in the () are the scores of the indicator.

For the CSP of each river, the St. Lawrence River, the Thames River and the Rhine River get the grade of “excellent” with very high scores of over 95.0 points; the Volga River, the Colorado River, the Mississippi River, the Yellow River, the Euphrates River, the Danube River and the Murray-Darling River get the grade of “good” with relatively high scores of over 85.0 points; the Yangtze River, with its CSP score being over 80 points, is graded as “medium high”; the Congo River gets its CPS score of 72.2 points and is graded as “medium”; and the Amazon River, the Ganges River, and the Nile River get relatively low scores of under 70 points and need to step up flood prevention and control.

5.3 Reliability of Water Resources (RWR)

The average score is 70.9 points and graded as “medium”, in general, on Reliability of Water Resources (RWR) (Figure 5.3). The high development intensity and insufficiency in water supply are pervasive issues for most rivers.

The rivers in North America performed high in RWR, while the African rivers performed relatively low. Among them, the St. Lawrence River in North America gets a score of 86.0 points, reaching the grade of "excellent", and both the Mississippi and Colorado River receive scores between 80 and 85 points, thus being graded as "medium high"; the Rhine River in Europe, the Murray-Darling River in Australia, and the Yangtze River and the Ganges in Asia all score between 70 and 80 points, belonging to the grade of "medium"; the Danube River and the Thames River in Europe, the Congo River in Africa, the Amazon River in South America, and the Volga River, the Euphrates River, and the Yellow River in Asia all receive scores between 60 and 70 points, and are graded as "medium low"; and the Nile River in Africa scores less than 60 points, falling into the grade of "quite poor".

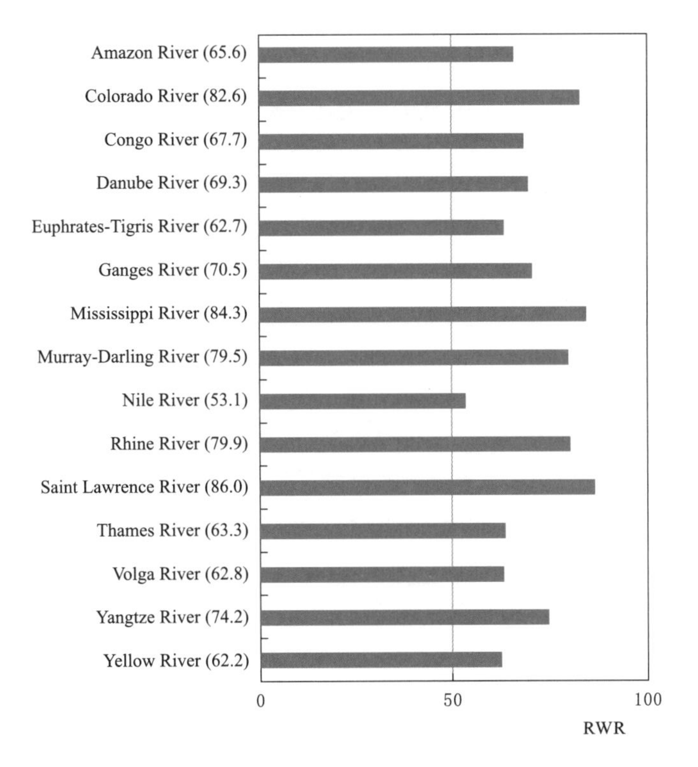

Figure 5.3 Evaluation results of RWR

Note: The scores in the () are the scores of the indicator.

5.4 Livability of Water Environment (LWE)

The rivers score an average 77.8 points and are graded as "medium high", in general, on the Livability of Water Environment (LWE) (Figure 5.4). It can be noted that how well the riverine countries develop and govern the rivers significantly sway the scores under the LWE. The scores of the LWR in economically developed areas are relatively high, and are lower in economically backward areas.

For the LWE, the Thames River, the Rhine River and the St. Lawrence River all get high scores of above 95 points, achieving the grade of "excellent"; the Yangtze River, the Danube River and the

Yellow River all score above 85 points, thus being graded as "good"; the scores of the Colorado River, the Mississippi River, the Murray-Darling River and the Volga River all exceed 80 points, being categorized as "medium high"; the Amazon River and the Euphrates River both score between 60 and 70 points, therefore being graded as "medium low"; however, the Ganges River, the Nile River and the Congo River, with their low scores between 30 and 60 points, are all graded as "quite poor".

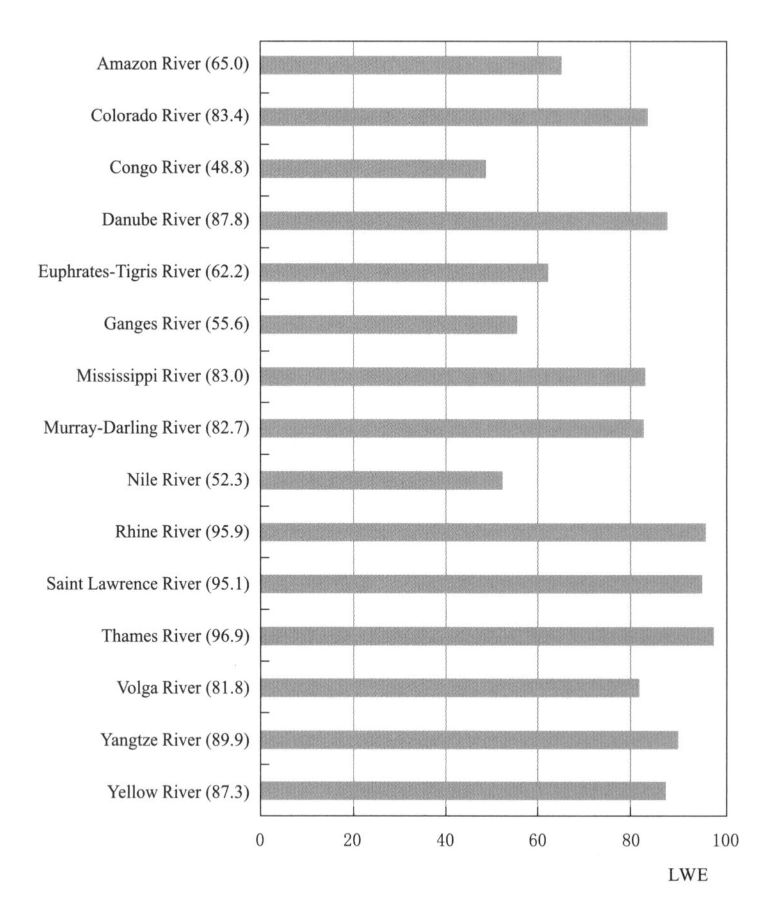

Figure 5.4 Evaluation results of LWE

Note: The scores in the () are the scores of the indicator.

5.5 Health of Aquatic Ecosystem (HAE)

The rivers score 71.8 points on average and are graded as "medium" on Health of Aquatic Ecosystem (HAE) in general. With the exception of the Congo and the Amazon, other rivers are subject to intensive development in their basins, resulting in reduced River Longitudinal Connectivity Index (LCI), and "quite poor" or "extremely poor" Variation of Eco-Hydrological Process (VIH). According to the proportion of endangered fish species in the basins, some of the rivers fall into "medium" or "poor" grades in Fish Endangered Index (FEI).

For the HAE, the Congo River and the Amazon River both get high scores between 85 and 95 points, thus being graded as "good"; the Yangtze River and the Volga River, are graded as "medium high" with both of their scores exceeding 80 points; the Rhine River, the Yellow River, the Colorado River, the Thames River and the Danube River all score between 70 and 80 points, therefore being graded as "medium"; the Mississippi River, the Ganges River, the Nile River and the Euphrates River all score between 60 and 70 points, and are graded as "medium low"; while the St. Lawrence River and the Murray-Darling River, with their low scores between 30 and 60 points, are graded as "quite poor".

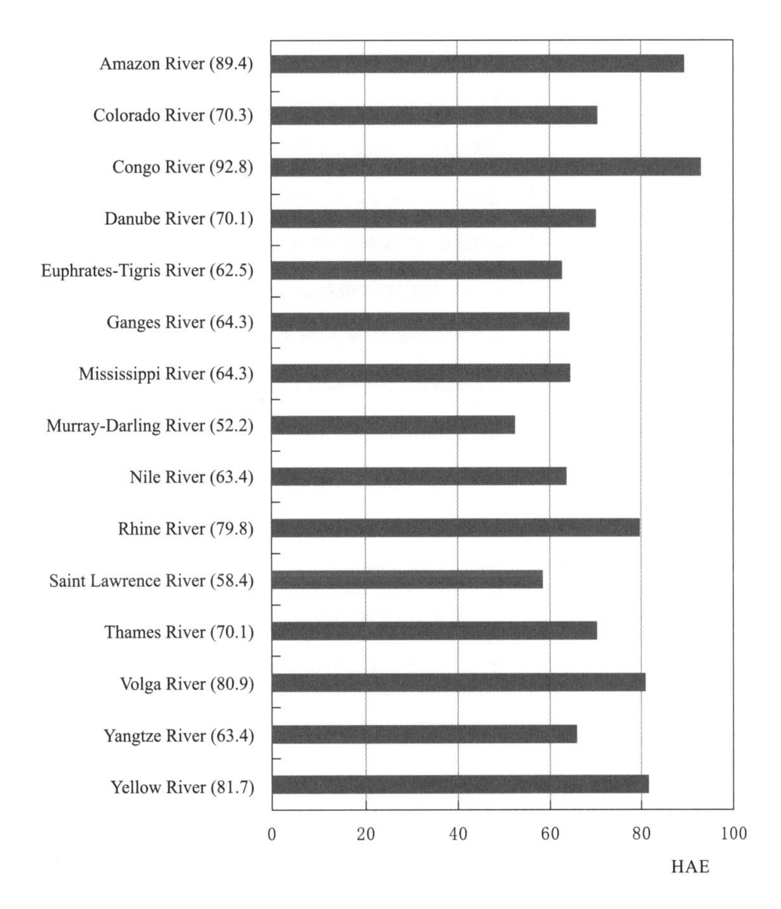

Figure 5.5 Evaluation results of HAE

Note: The scores in the () are the scores of the indicator.

5.6 Prosperity of Water Culture (PWC)

The rivers score an average of 77.4 points and are graded as "medium" or above on the Prosperity of Water Culture (PWC) in general, exhibiting rich water cultural legacies. However, there is room for improvement in the Modern Water Culture Creation and Innovation (MCI), and the Public Awareness and Involvement in Water Governance (PAG).

For the PWC, the Danube River and the Thames River both score between 85 and 95 points, and are graded as "good"; the Nile River, the St. Lawrence River and the Yellow River score between 80 and 85 points, fitting into the grade of "medium high"; the other rivers including the Amazon River, the Colorado River, the Euphrates River, the Ganges River, the Mississippi River, the Murray-Darling River, the Rhine River, the Volga River and the Yangtze River all score between 70 and 80 points, lying in the grade of "medium"; while the Congo River, getting the lowest score of 67.5 points, falls into the grade of "medium poor".

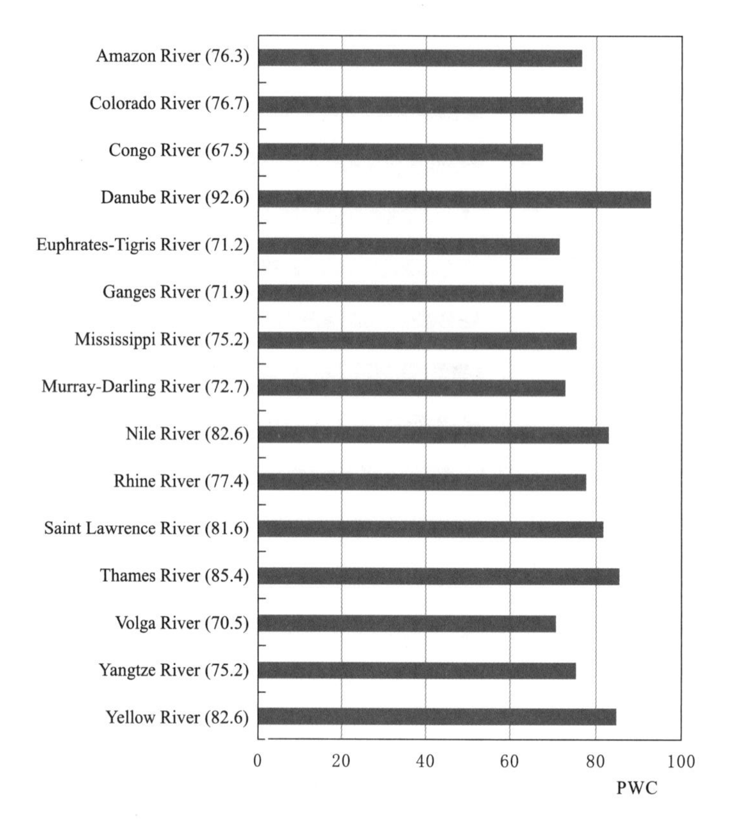

Figure 5.6 Evaluation results of PWC

Note: The scores in the () are the scores of the indicator.

Appendix: Sources of Information

1. World Bank Database.
2. Website of Food and Agriculture Organization of the United Nations (FAO).
3. University of Notre Dame, Global Adaptation Index(ND-GAIN).
4. Database of National Oceanic and Atmospheric Administration (NOAA).
5. Database of Open Street Map.
6. Database of Google Earth Engine.
7. Emergency Event Database (EM-DAT) of the World Health Organization(WHO) and the Belgian Government.
8. Goal 6 of the UN Sustainable Development Goals (SDGs).
9. EEA and EU reports on water resources assessment.
10. Water Quality Data of the United States Geological Survey (USGS).
11. The Global Freshwater Quality Database GEMStat.
12. FishBase.
13. Red List of Threatened Species of International Union for Conservation of Nature (IUCN).
14. The European Space Agency (ESA) WorldCover 10m 2020.
15. Global Reservoir and Dam (GRanD) Database.
16. Grill G. et al. Mapping the world's free-flowing rivers. *Nature*. 2019,569: 215–221.
17. Global Rivers and Sedimentation Databases of International Research and Training Center on Erosion and Sedimentation (IRTCES),UNESCO.
18. China River Happiness Report 2020.
19. China Water Resources Bulletin 2020.
20. China Bulletin on Flood and Drought Disaster Mitigation 2011-2020.
21. China River Sediment Bulletin.